AF801639

Manhattan Review

Test Prep & Admissions Consulting

Turbocharge Your GMAT: Quantitative Reasoning Question Bank

Published on August 20th, 2024

Version 416

- ☐ *Mapped according to the GMAT Focus Edition*
- ☐ *200+ GMAT-like practice questions*
- ☐ *Comprehensive, Intuitive, and graphical explanations of concepts*
  - *Great collection of 700+ level questions*
  - *Ample questions with Alternate Approaches*
- ☐ *Detailed explanations of questions*

www.manhattanreview.com

Copyright and Terms of Use

Copyright and Trademark

Terms of Use

No Warranties

Limitation on Liability

International Standard Book Number (ISBN): 978-1-62926-113-3)

Published on August 20th, 2024.

Manhattan Review, 2 Park Avenue, Suite 2010, New York, NY 10016.
Phone: +1 (212) 316-2000. E-Mail: info@manhattanreview.com. Web: www.manhattanreview.com

About the Turbocharge your GMAT Series (Focus Edition)

The Turbocharge Your GMAT Series is carefully designed to be clear, comprehensive, and content-driven. Long regarded as the gold standard in GMAT prep worldwide, Manhattan Review's GMAT prep books offer professional GMAT instruction for dramatic score improvement. Now in its updated 7th edition, the full series is designed to provide GMAT Focus test-takers with complete guidance for highly successful outcomes. As many students have discovered, Manhattan Review's GMAT books break down the different test sections in a coherent, concise, and accessible manner. We delve deeply into the content of every single testing area and zero in on exactly what you need to know to raise your score. The full series covers concepts in mathematics from the most basic through the most advanced levels, making them a great study resource for all stages of GMAT preparation. Students who work through all of our books benefit from a substantial boost to their GMAT knowledge and develop a thorough and strategic approach to taking the GMAT.

- ☐ **GMAT Focus Math Essentials (ISBN: 978-1-62926-108-9)**
- ☐ **GMAT Focus Quantitative Reasoning Guide (ISBN: 978-1-62926-109-6)**
- ☐ **GMAT Focus Reading Comprehension Guide (ISBN: 978-1-62926-110-2)**
- ☐ **GMAT Focus Critical Reasoning Guide (ISBN: 978-1-62926-111-9)**
- ☐ **GMAT Focus Data Insights Guide (ISBN: 978-1-62926-112-6)**
- ■ **GMAT Focus Quantitative Reasoning Question Bank (ISBN: 978-1-62926-113-3)**
- ☐ **GMAC Business Writing Assessment Guide (ISBN: 978-1-62926-114-0)**

About the Company

Manhattan Review's origin can be traced directly back to an Ivy League MBA classroom in 1999. While teaching advanced quantitative subjects to MBAs at Columbia Business School in New York City, Professor Dr. Joern Meissner developed a reputation for explaining complicated concepts in an understandable way. Remembering their own less-than-optimal experiences preparing for the GMAT, Prof. Meissner's students challenged him to assist their friends, who were frustrated with conventional GMAT preparation options. In response, Prof. Meissner created original lectures that focused on presenting GMAT content in a simplified and intelligible manner, a method vastly different from the voluminous memorization and so-called tricks commonly offered by others. The new approach immediately proved highly popular with GMAT students, inspiring the birth of Manhattan Review.

Since its founding, Manhattan Review has grown into a multi-national educational services firm, focusing on GMAT preparation, MBA admissions consulting, and application advisory services, with thousands of highly satisfied students all over the world. The original lectures have been continuously expanded and updated by the Manhattan Review team, an enthusiastic group of master GMAT professionals and senior academics. Our team ensures that Manhattan Review offers the most time-efficient and cost-effective preparation available for the GMAT. Please visit www.ManhattanReview.com for further details.

About the Founder

Professor Dr. Joern Meissner has more than 25 years of teaching experience at the graduate and undergraduate levels. He is the founder of Manhattan Review, a worldwide leader in test prep services, and he created the original lectures for its first GMAT preparation class. Prof. Meissner is a graduate of Columbia Business School in New York City, where he received a PhD in Management Science. He has since served on the faculties of prestigious business schools in the United Kingdom and Germany. He is a recognized authority in the areas of supply chain management, logistics, and pricing strategy. Prof. Meissner thoroughly enjoys his research, but he believes that grasping an idea is only half of the fun. Conveying knowledge to others is even more fulfilling. This philosophy was crucial to the establishment of Manhattan Review, and remains its most cherished principle.

The Advantages of Using Manhattan Review

- **Time efficiency and cost effectiveness.**
  - For most people, the most limiting factor of test preparation is time.
  - It takes significantly more teaching experience to prepare a student in less time.
  - Our test preparation approach is tailored for busy professionals. We will teach you what you need to know in the least amount of time.
- **Our high-quality and dedicated instructors are committed to helping every student reach her/his goals.**

International Phone Numbers and Official Manhattan Review Websites

Manhattan Headquarters	+1-212-316-2000	www.manhattanreview.com
USA & Canada	+1-800-246-4600	www.manhattanreview.com
Argentina	+1-212-316-2000	www.review.com.ar
Australia	+61-3-9001-6618	www.manhattanreview.com
Austria	+43-720-115-549	www.review.at
Belgium	+32-2-808-5163	www.manhattanreview.be
Brazil	+1-212-316-2000	www.manhattanreview.com.br
Chile	+1-212-316-2000	www.manhattanreview.cl
China	+86-20-2910-1913	www.manhattanreview.cn
Czech Republic	+1-212-316-2000	www.review.cz
France	+33-1-8488-4204	www.review.fr
Germany	+49-89-3803-8856	www.review.de
Greece	+1-212-316-2000	www.review.com.gr
Hong Kong	+852-5808-2704	www.review.hk
Hungary	+1-212-316-2000	www.review.co.hu
India	+1-212-316-2000	www.review.in
Indonesia	+1-212-316-2000	www.manhattanreview.id
Ireland	+1-212-316-2000	www.gmat.ie
Italy	+39-06-9338-7617	www.manhattanreview.it
Japan	+81-3-4589-5125	www.manhattanreview.jp
Malaysia	+1-212-316-2000	www.review.my
Mexico	+1-212-316-2000	www.manhattanreview.mx
Netherlands	+31-20-808-4399	www.manhattanreview.nl
New Zealand	+1-212-316-2000	www.review.co.nz
Philippines	+1-212-316-2000	www.review.ph
Poland	+1-212-316-2000	www.review.pl
Portugal	+1-212-316-2000	www.review.pt
Qatar	+1-212-316-2000	www.review.qa
Russia	+1-212-316-2000	www.manhattanreview.ru
Singapore	+65-3158-2571	www.gmat.sg
South Africa	+1-212-316-2000	www.manhattanreview.co.za
South Korea	+1-212-316-2000	www.manhattanreview.kr
Sweden	+1-212-316-2000	www.gmat.se
Spain	+34-911-876-504	www.review.es
Switzerland	+41-435-080-991	www.review.ch
Taiwan	+1-212-316-2000	www.gmat.tw
Thailand	+66-6-0003-5529	www.manhattanreview.com
Turkey	+1-212-316-2000	www.review.com.tr
United Arab Emirates	+1-212-316-2000	www.manhattanreview.ae
United Kingdom	+44-20-7060-9800	www.manhattanreview.co.uk
Rest of World	+1-212-316-2000	www.manhattanreview.com

Contents

Chapter 1

Welcome

Dear Students,

At Manhattan Review, we go above and beyond to supercharge your standardized test preparation with the ultimate educational content. Our dynamic company is committed to improvement, which is why we relentlessly update our innovative educational approaches to bring you the most useful and comprehensive GMAT preparation resources available. Ready to master the GMAT Focus? This book will show you how to achieve your highest test score yet.

Navigating through the maze of GMAT test-prep resources can seem daunting. Your goal is to partner with the organization that best fits your learning- and test-taking needs, allowing you to feel confident and prepared on the day of your exam. At Manhattan Review, your success is our motivation, and we can't wait to empower your test preparation experience.

One of the most effective ways to strengthen your GMAT preparation is by completing practice questions, as the more familiar you are with the way questions are asked, the easier it is to arrive at the correct answer. At Manhattan Review, our experts spend a significant amount of time collecting and creating the most representative and challenging Quantitative Reasoning questions. With a potent blend of text and dynamic graphics, we ensure you aren't simply memorizing information but truly learning how to break down and solve even the most challenging problems. We teach you to think like a test-maker so that nothing will come as a surprise on the day of your exam.

Inside this book, you'll encounter a wide and vivid spectrum of problems, each offering a novel perspective important to your GMAT success. The alternate solutions to particularly intricate questions are especially insightful. We've also crafted smart traps mimicking actual GMAT challenges, coupled with insightful guidance to lead you to the right answer. Embark on an enlightening journey through 200+ questions in this power-packed volume.

Manhattan Review's collection of Quantitative Reasoning questions leaves no question unexplored while offering a well-rounded and expansive study companion. And guess what? We're always here for your queries! Just drop a line at *info@manhattanreview.com.*

Happy Learning!

Professor Dr. Joern Meissner
& The Manhattan Review Team

Chapter 2

Problem Solving Question Bank

2.1 Number properties

1. $99996^2 - 4^2 =$

(A) $10^{10} - 8$

(B) $(10^5 - 8)^2$

(C) $10^4(10^5 - 8)$

(D) $10^5(10^4 - 8)$

(E) $10^5(10^5 - 8)$

2. If 5^a is a factor of $n!$, and the greatest integer value of a is 6, what is the largest possible value of b such that 7^b is a factor of the largest possible value of $n!$?

(A) 2

(B) 3

(C) 4

(D) 5

(E) 6

3. If $a = 0.999$, $b = (0.999)^2$ and $c = \sqrt{0.999}$, which of the following is the correct order of a, b and c?

(A) $a < b < c$

(B) $a < c < b$

(C) $b < c < a$

(D) $b < a < c$

(E) $c < a < b$

4. If p is the product of the reciprocals of integers from 150 to 250, inclusive, and q is the product of the reciprocals of integers from 150 to 251, inclusive, what is the value of $(p^{-1} + q^{-1})$ in terms of p?

(A) $\frac{p}{(251)^2}$

(B) $251 \times 252 \times p$

(C) $252p$

(D) $\frac{252}{p}$

(E) $251 \times 252 \times p^2$

5. If x is the sum of all integers from 51 to 100, inclusive, what is the value of x?

(A) 3,624

(B) 3,625

(C) 3,675

(D) 3,725

(E) 3,775

6. If x is the sum of the reciprocals of the consecutive integers from 51 to 60, inclusive and y is the sum of the reciprocals of the consecutive integers from 61 to 70, inclusive, which of the following is correct?

I. $\frac{1}{x} > 6$

II. $\frac{1}{y} > 7$

III. $\frac{1}{y} > \frac{1}{x}$

(A) Only I
(B) Only II
(C) Only III
(D) Only II and III
(E) I, II and III

7. A number $4p25q$ is divisible by 4 and 9; where p and q are the thousands and units digits, respectively. What is the minimum value of $\frac{p}{q}$?

(A) $\frac{1}{8}$
(B) $\frac{1}{7}$
(C) $\frac{1}{6}$
(D) $\frac{2}{5}$
(E) $\frac{5}{2}$

8. If a and b are real numbers such that a percent of $(a - 2b)$ when added to b percent of b, the value obtained is 0, then which of the following statements must be true?

I. $a = b$
II. $a + b = 0$
III. $a - b = 1$

(A) Only I
(B) Only II
(C) Only III
(D) Only I and III
(E) Only II and III

9. A set is such that if m is in the set, $(m^2 + 3)$ is also in the set. If -1 is in the set, which of the following is also in the set?

I. -2
II. 4
III. 19

(A) Only I
(B) Only II

(C) Only I and II

(D) Only II and III

(E) I, II and III

10. A sequence $t_1, t_2, t_3, \ldots t_n$ is such that $t_2 = 5$ and $t_{n+1} = 2t_n - 1$ for $n \geq 1$, then what is the value of $t_{10} - t_9$?

(A) 2^9

(B) $2^{10} + 1$

(C) 1

(D) 2

(E) 4

11. If m, n, p, and q are distinct positive integers, and all are greater than 1 such that $mnpq = 660$, how many possible combination of values exist for m, n, p, and q?

(A) Two

(B) Three

(C) Four

(D) Five

(E) Seven

12. If $x = 125$ and $y = 216$, what is the value of $\left(\sqrt{x^{2/3} + y^{2/3} + 2(xy)^{1/3}} + \sqrt{x^{2/3} + y^{2/3} - 2(xy)^{1/3}}\right)$?

(A) 0

(B) 5

(C) 6

(D) 10

(E) 12

13. If $m \geq 0.9$, which of the following is a possible value of $\left(\frac{1}{\sqrt{m}}\right)$?

(A) 1.01

(B) 1.12

(C) 1.13

(D) 1.35

(E) 1.50

14. Given that a and b are positive integers and $a = \frac{b^3}{90}$, which of the following must be an integer?

I. $\frac{a}{45}$

II. $\frac{a}{90}$

III. $\frac{a}{300}$

(A) Only I

(B) Only II
(C) Only III
(D) Only I and II
(E) I, II and III

15. The addition operation shown below shows the sum of the two-digit positive integers XY and YX. If X, Y, and Z are different digits, what is the value of the integer Z?

$$\begin{array}{r} X\ Y \\ +\ Y\ X \\ \hline X\ X\ Z \\ \hline \end{array}$$

(A) 8
(B) 7
(C) 2
(D) 1
(E) 0

16. Suzy saves \$20 per month. In each of the 30 months, she saved \$20 more than he saved in the previous month. What is the total amount she saved during the 30-month period?

(A) \$3,600
(B) \$4,800
(C) \$6,000
(D) \$9,300
(E) \$12,000

17. If a sequence of numbers $t_1, t_2, \ldots t_n$ is such that $t_1 = 0,\ t_2 = 2$ and $t_n = t_{(n+1)} + 2t_{(n-1)}$ for $n \geq 1$, what is the value of t_4 ?

(A) -2
(B) 0
(C) 4
(D) 6
(E) 8

18. If n is an integer such that $n > 9$, which of the following could be the remainder when $(2 + 2^2 + 2^3 + 2^4 + \ldots + 2^n)$ is divided by 3?

I. 0
II. 1
III. 2

(A) Only I
(B) Only II

(C) Only III

(D) Only I and III

(E) I, II and III

2.2 Percents

19. A machine can be repaired for $1,200 and will last for one year, while a new machine would cost $2,800 and will last for two years. The average cost per year of the new machine is what percent greater than the cost of repairing the current machine?

(A) 7%

(B) 14.28%

(C) 16.67%

(D) 18.83%

(E) 20%

20. An item is subject to a sales tax of 10 percent on the part of the price that is greater than $200. If a customer paid a sales tax of $10 on the item, what was the price of the item?

(A) $200

(B) $250

(C) $300

(D) $360

(E) $400

21. Item A is subject to a sales tax rate of $0.54 per $25. What is the sales tax rate, as a percent, for item B that is subject to four times as much as the sales tax rate for item A?

(A) 216%

(B) 86.4%

(C) 8.64%

(D) 2.16%

(E) 0.135%

22. In the last lap, cyclist P increases his speed from 10 miles per hour to 25 miles per hour, while cyclist Q increases his speed from 8 miles per hour to 24 miles per hour. By what percent is the percent increase in speed of cyclist Q greater than the percent increase in speed of cyclist P?

(A) 33.33%

(B) 50%

(C) 66.67%

(D) 75%

(E) 100%

23. In a certain year, Carrier X traveled 101,098 kilometers and consumed 9,890 liters of diesel fuel, while in the same year, Carrier Y traveled 203,000 kilometers and consumed 24,896 liters of diesel fuel. Fuel mileage is defined as kilometers per liter of fuel. The mileage of Carrier X is approximately what percent greater or lesser than that of Carrier Y?

(A) 20%

(B) 25%

(C) 33.33%

(D) 37.50%

(E) 40%

24. The list price of a bicycle is \$456. The trader first decreased the price by 25 percent and then increased it by 25 percent.†The final price of the bicycle is what percent of the list price?

(A) 0%

(B) 50%

(C) 66.67%

(D) 93.75%

(E) 100%

25. To prepare a certain color, a paint dealer mixes 3.4 liters of red color into a base that is 68 liters. The paint manufacturer recommends mixing 0.7 liters per 10 liters of base to make that color. By what percent should the dealer increase the quantity of the paint to match the recommendation?

(A) 10%

(B) 33.33%

(C) 40%

(D) 66.66%

(E) 72%

26. A Business Processing Outsourcing unit recruits 200 employees. Each of them is paid \$7.50 per hour for the first 44 hours worked during a week and $1\frac{1}{3}$ times that rate for hours worked in excess of 44 hours. What was the total remuneration of the employees for a week in which 30 percent of them worked 30 hours, 40 percent worked 44 hours, and the rest worked 50 hours?

(A) \$25,000

(B) \$60,600

(C) \$63,300

(D) \$70,000

(E) \$73,400

27. A retail company earned \$5 million as commission on the first \$35 million in sales and then \$11 million as commission on the next \$121 million in sales. By what percent did the ratio of commissions to sales decrease from the first \$35 million in sales to the next \$121 million in sales?

(A) 11.11%

(B) 22.22%

(C) 36.36%

(D) 44.44%

(E) 50%

28. A sales representative earned 8 percent commission on the amount of sales up to \$20,000, inclusive, and x percent commission on the amount of sales above \$20,000. If the representative earned a total commission of \$2,000 on sales of \$24,000, what was the value of x ?

(A) 4
(B) 6
(C) 8
(D) 10
(E) 12

29. A trader buys a batch of 120,000 computer chips for $3,600,000. He sells two-fifths of the computer chips, each at 25 percent above the cost per computer chip. He sells the remaining chips at a price per chip equal to 25 percent less than the cost per chip. What was the percent profit or loss on the batch of the chips?

(A) Loss of 1%
(B) Loss of 5%
(C) Loss of 7.50%
(D) Profit of 10%
(E) Profit of 22.22%

30. With the 20% increase in the price of milk, a housewife can buy 5 liters of less quantity for $60 than before the increase. What was the initial price per liter of milk?

(A) $2.00
(B) $2.50
(C) $2.75
(D) $3.00
(E) $3.50

31. A company was approved to spend a certain sum of money for a year. It spent one-fourth of the sum during the first quarter and one-sixth of the remainder during the second quarter. By what percent is the sum of money that was left at the beginning of the third quarter more than the sum spent in the two quarters?

(A) 10%
(B) 22.22%
(C) 33.33%
(D) 66.66%
(E) 133.33%

32. David and Suzy each spent $450 in 2013. In 2014, David spent 10 percent more than he did in 2013, and they together spent $600. Approximately, by what percent less did Suzy spend in 2014 than she did in 2013?

(A) 23%
(B) 66%
(C) 77%
(D) 80%
(E) 83%

33. On day 1, a shopkeeper increases the price of an item by k%, and on day 2, he decreases the price by k%. By the end of day 2, the price of the item drops by \$1. On day 3, he again increases the price by k%, and on day 4, he again decreases the price by k%. If, at the end of day 4, the price of the item is \$398, what is the approximate initial price of the item?

(A) \$325
(B) \$350
(C) \$375
(D) \$400
(E) \$450

34. Last year, \$$y$ could buy x number of items. If the average cost of the item increased by 20 percent this year, then the number of items that can be bought with \$$3y$ equals

(A) x
(B) $1.50x$
(C) $2.50x$
(D) $3x$
(E) $3.50x$

35. A solution consists of 30 percent water by weight. After boiling the solution for 15 minutes, 70 percent of the water, by weight, evaporated. There is no weight loss for the other part of the solution. What percent of the solution's total remaining weight consists of the remaining water?

(A) $\frac{500}{69}\%$
(B) $\frac{600}{69}\%$
(C) $\frac{700}{79}\%$
(D) $\frac{900}{79}\%$
(E) $\frac{100}{69}\%$

36. A mixed juice contains, by volume, 25 percent banana pulp and 75 percent papaya pulp. If this mixed juice costs 20 percent more than an equal quantity of only banana pulp, by what percent is papaya pulp more expensive than banana pulp?

(A) 22.22%
(B) 26.67%
(C) 28%
(D) 30%
(E) 33.33%

37. At a lab, bacteria P multiplies itself every 18 days, while bacteria Q multiplies itself every 15 days. Approximately by what percent is the number of times bacteria Q multiplies itself greater than the number of times bacteria P multiplies itself in a 3-year period? Assume that one year equals 365 days.

(A) 12%

(B) 16%

(C) 20%

(D) 22%

(E) 33%

38. Jack purchased a phone for \$1,500 and paid tax at the rate of 5 percent, while Tom purchased a phone for \$1,200 and paid tax at the rate of 15 percent. The total amount Tom paid was what percent less than the total amount Jack paid?

(A) 5%

(B) 7%

(C) 9%

(D) 12%

(E) 15%

39. In a class, 65 percent of the boys and 78 percent of the girls play basketball. If 72 percent of all the students play basketball, what is the ratio of the number of girls to the number of boys?

(A) $\frac{4}{3}$

(B) $\frac{7}{6}$

(C) $\frac{8}{7}$

(D) $\frac{9}{8}$

(E) $\frac{13}{11}$

40. In a stadium, the Royal Challengers team had a support of 24,500 spectators from natives and 10 percent of spectators from other than natives. If S is the total number of spectators in the stadium and 40 percent belonged to natives, which of the following represents the number of supporters for the Royal Challengers team?

(A) $0.6S + 12,250$

(B) $0.28S + 12,250$

(C) $0.28S + 24,500$

(D) $0.06S + 24,500$

(E) $0.6S + 24,500$

41. In a school, 40 percent of the students study science, and 60 percent of them go to special classes after school. If 30 percent of the students at the school go to special classes, what percent of the total students who do not study science go to special classes?

(A) 6%

(B) 12%

(C) 15%

(D) 24%

(E) 27%

42. In a class at a school, there were 40 percent boys. If some students were transferred to a new section and 30 percent of the transferred students were boys, what was the ratio of the transfer rate for the boys to the transfer rate for the girls?

(A) 1 : 4

(B) 2 : 7

(C) 4 : 9

(D) 9 : 14

(E) 9 : 16

43. At the beginning of a year, a car was valued $\left(\frac{5}{7}\right)^{\text{th}}$ of the original price, and at the end of the year, it was value $\left(\frac{3}{5}\right)^{\text{th}}$ of the original price. By what percent did the value of the car decrease in the year?

(A) 11.11%

(B) 16%

(C) 17.50%

(D) 19%

(E) 22.22%

44. A salesman is offered either a 5 percent commission on his monthly sales, in dollars, and a monthly bonus of $500, or a 7 percent commission on his monthly sales with no bonus. At what sales, in dollars, will both offers give him the same remuneration?

(A) $22,500

(B) $25,000

(C) $32,500

(D) $35,000

(E) $40,000

45. In the beginning of the year, 35 percent of company X's 120 customers were retailers, and after the 24-month period, 25 percent of its 240 customers were retailers. What was the simple annual percent growth rate in the number of retailers?

(A) 14.28%

(B) 21.43%

(C) 24.0%

(D) 30.0%

(E) 37.25%

46. Which of the following gives the highest overall percent increase, if, in each case, the second percent increase is applied to the value obtained after applying the first percent increase?

(A) 10 percent increase followed by 50 percent increase

(B) 25 percent increase followed by 35 percent increase

(C) 30 percent increase followed by 30 percent increase

(D) 40 percent increase followed by 20 percent increase

(E) 45 percent increase followed by 15 percent increase

47. In a school, 70 percent of students are boys, and the rest are girls. In a prefect election, 30 percent of boys and 70 percent of girls voted for John. What percent of the total students voted for John?

(A) 37%
(B) 42%
(C) 50%
(D) 58%
(E) 66%

48. According to the table given below, a state has a total of 23,000 companies from seven regions. By what percent of the total number of companies in the region is the number of companies in Region S greater than the number of companies in Region R?

Region-wise distribution of companies in the state	
Regions	**No. of companies**
Region P	2,345
Region Q	3,456
Region R	3,421
Region S	5,721
Region T	3,445
Region U	80
Region V	4,532

(A) 5%
(B) 10%
(C) 17.5%
(D) 22.5%
(E) 33.33%

2.3 Profit & Loss

49. A shopkeeper could sell only $\left(\frac{4}{5}\right)^{\text{th}}$ of the stock at the rate of \$3 per item. If 100 items were unsold, what was the total amount he received from the sale?

(A) \$240
(B) \$1,200
(C) \$1,250
(D) \$1,300
(E) \$1,500

50. A trader bought 900 cartons of a certain icecream brand at a cost of \$20 per carton. If he sold $\left(\frac{2}{3}\right)^{\text{rd}}$ of the cartons for one and a quarter times their cost price and sold the remaining cartons at a loss of 20 percent of their cost price, what was the trader's gross profit on the total sale?

(A) \$1,800
(B) \$2,400
(C) \$2,700
(D) \$3,000
(E) \$3,200

51. A dealer sells only two brands of bicycles, brand A and brand B. The selling price of a brand A bicycle is \$150, which is 60 percent of the selling price of a brand B bicycle. If the dealer sells 100 pieces of bicycles, and $\left(\frac{3}{5}\right)^{\text{th}}$ of which are brand B, what is the dealer's total sales, in dollars, from the sale of bicycles?

(A) \$15,000
(B) \$16,000
(C) \$18,000
(D) \$21,000
(E) \$22,000

52. A trader bought a consignment at a purchase price of \$800 and sold it for 20% less than the marked price. If the trader made a profit equivalent to 30% of the purchase price, what is the marked price of the consignment?

(A) \$1,000
(B) \$1,200
(C) \$1,300
(D) \$1,350
(E) \$1,500

53. A small textile company buys a few machines to stitch garments, costing a total of \$10,000. The per-unit cost of each garment is \$2.50, and is sold for \$4.50. How many units of the garments must be sold to achieve break-even (a phenomenon when all the investment and production costs are recovered by the sales revenue)?

(A) 2,000
(B) 3,500
(C) 4,500
(D) 5,000
(E) 6,000

54. A broker sold a house with a gross margin of 20 percent on the cost of the house. If the selling price of the house were increased by $10,000, it would yield a gross margin of 30 percent of the cost of the house. What was the original selling price of the house?

(A) $90,000
(B) $100,000
(C) $120,000
(D) $140,000
(E) $150,000

55. A television assembler pays its contractors $20 each for the first 100 assembled sets and $15 for each additional set. If 600 television sets were assembled and the assembler invoiced the manufactures $25 for each set, what was the assembler's gross profit in dollars?

(A) $3,750
(B) $4,500
(C) $5,500
(D) $6,000
(E) $7,000

56. A merchant's gross profit on item A was 10 percent of its cost. If the merchant increased its selling price from $99 to $117, keeping its cost the same, the merchant's profit on item A after the price increase was what percent of the cost of item A?

(A) 20%
(B) 21%
(C) 24%
(D) 27%
(E) 30%

57. A merchant bought 2,400 fans for $30 each. He sold 60 percent of the fans for $40 each, and the rest for $35 each. What was the merchant's average profit per fan?

(A) $6
(B) $8
(C) $9
(D) $10
(E) $12

58. A trader sold an item at a m percent profit on cost after giving a m percent discount on the list price. Had he sold the item at a $(m + 15)$ percent discount on the list price, his profit would have been $(m - 20)$ percent on cost. What would have been his percent profit had he sold the item without giving any discount?

(A) 5.0%
(B) 10.0%
(C) 25.0%
(D) 33.3%
(E) 38.0%

59. A merchant sold 800 units of bedsheets for \$8 each and 900 units of bedsheets for \$5 each. If the merchant's cost of producing each unit of bedsheet was \$6, what was the merchant's profit or loss on the sale of 1,700 bedsheets?

(A) Loss of \$700
(B) Loss of \$300
(C) No profit or loss
(D) Profit of \$300
(E) Profit of \$700

60. The sales revenue from book sales in 2015 was 10% less than that in 2014, and the sales revenue from stationary sales in 2015 was 6% more than that in 2014. If total sales revenues from book sales and stationary sales in 2015 were 2% more than those in 2014, what is the ratio of sales revenue from book sales in 2014 to sales revenue from stationary sales in 2014?

(A) 1 : 3
(B) 2 : 3
(C) 3 : 4
(D) 4 : 5
(E) 5 : 6

61. A trader's profit in 2002 was 20 percent higher than in 2001, and its profit in 2003 was 25 percent higher than its profit in 2002. The company's profit in 2003 was what percent higher than its profit in 2001?

(A) 5%
(B) 45%
(C) 46%
(D) 48%
(E) 50%

2.4 Averages

62. Milton School has a student-to-teacher ratio of 25 to 2. The average (arithmetic mean) annual salary for teachers is \$42,000. If the school pays a total of \$3,780,000 in annual salaries to its teachers, how many students does the school have?

(A) 900
(B) 1,000
(C) 1,125
(D) 1,230
(E) 1,500

63. The average (arithmetic mean) annual salary of the employees of a company was \$70,000. If the male employees' annual salary average was \$65,000 and that of female employees' annual salary was \$80,000, what could be the number of male employees and female employees, respectively, in the company?

(A) 6; 7
(B) 7; 15
(C) 7; 14
(D) 14; 7
(E) 15; 7

64. A class consists of 40 students and is divided into two sections. In Section A, the average score in a test was 85. In Section B, the average score in the test was 80. If the average score of the class in the test was 82, how many students are in Section A?

(A) 12
(B) 14
(C) 16
(D) 20
(E) 22

65. A juice manufacturer has 1,200 liters of mango pulp in stock, 25 percent of which is water. If the manufacturer adds another 400 liters of mango pulp, of which 20 percent is water, what percent, by volume, of the manufacturer's mango pulp contains water?

(A) 21.50%
(B) 23.75%
(C) 33.33%
(D) 35.00%
(E) 37.50%

66. A class has four sections: P, Q, R, and S, and the average weights of the students in the sections are 45 lb, 50 lb, 55 lb, and 65 lb, respectively. What is the maximum possible number of students in Section R if there are 40 students in the class and the average weight of all students o the class is 55 lb?

(A) 18

(B) 20

(C) 25

(D) 35

(E) 37

67. If set N consists of odd numbers of consecutive integers, starting with 1, what is the difference between the average of the odd integers and the average of the even integers in set N?

(A) -1

(B) 0

(C) $\frac{1}{2}$

(D) 1

(E) 2

68. The average of nine numbers is 25. The average of the first five numbers is 20, and that of the last five is 32. What is the value of the fifth number?

(A) 30

(B) 32

(C) 35

(D) 36

(E) 38

69. Box X and Box Y each contain many yellow and green balls. All of the green balls have the same radius. The radius of each green ball is 4 inches less than the average radius of the balls in Box X and 2 inches greater than the average radius of the balls in Box Y. What is the difference between the average (arithmetic mean) radius, in inches, of the balls in Box X and the balls in Box Y?

(A) 4

(B) 6

(C) 7

(D) 8

(E) 10

70. A certain company has 60 employees. The average (arithmetic mean) salary of 10 of the employees is \$35,000; the average salary of 35 other employees is \$30,000; and the average salary of the remaining 15 employees is \$60,000. What is the average salary of the 60 employees at the company?

(A) \$32,500

(B) \$38,333

(C) \$39,500

(D) \$40,000

(E) \$42,222

71. At a certain stationery shop, the price of a pencil is 20 cents, and the price of an eraser is 30 cents. A boy buys a total of 20 pencils and erasers from the shop, and the average (arithmetic mean) price of the 20 pieces comes to 28 cents. How many erasers must the boy return so that the average price of the pieces that he buys is 26 cents?

(A) 2
(B) 4
(C) 6
(D) 8
(E) 10

72. A student's average (arithmetic mean) test score on four tests is 78. If each test is scored out of 100, which of the following can be the student's score on the fifth test so that the student's average score on five tests increases by an integer value?

(A) 82
(B) 87
(C) 89
(D) 93
(E) 95

73. An instructor gave the same test to three groups: P, Q, and R. The average (arithmetic mean) scores for the three groups were 64, 84, and 72, respectively. The ratio of the numbers of candidates in the P, Q, and R groups was 3 : 5 : 4, respectively. What was the average score for the three groups combined?

(A) 72
(B) 75
(C) 77
(D) 78
(E) 80

74. A fitness club has 50 male and 20 female members. The average (arithmetic mean) age of all of the members is 23 years. If the average age of the male members was 20 years, which of the following is the average age, in years, of the female members?

(A) 30.50
(B) 31.50
(C) 32.50
(D) 33.00
(E) 34.50

75. Following is a modified question of the above.

A fitness club has 50 male and 20 female members. The average (arithmetic mean) age of all of the members is 23.89 years. If the average age of the male members was 20.89 years, which of the following is the average age, in years, of the female members?

(A) 29.75

(B) 30.50
(C) 31.39
(D) 34.39
(E) 37.50

2.5 Ratio & Proportion

76. The total cost of manufacturing metal bearings incurs a fixed cost of \$25,000 and a variable expense, which depends on the number of bearings manufactured. If for 50,000 bearings the total cost is \$100,000, what is the total cost for 100,000 bearings?

(A) \$125,000
(B) \$150,000
(C) \$175,000
(D) \$200,000
(E) \$275,000

77. A beaker was filled with a 40 liters of mixture. The mixture has water and a liquid chemical in the ratio of 3 : 5, respectively. If each day, for a 10-day period, 2 percent of the initial quantity of water and 5 percent of the initial quantity of liquid chemical evaporated, what percent of the original amount of mixture evaporated during this period?

(A) 22.22%
(B) 33.33%
(C) 38.75%
(D) 44.44%
(E) 58.33%

78. In Ghazal's doll collection, $\left(\frac{3}{5}\right)^{\text{th}}$ of the dolls are Barbie dolls, and $\left(\frac{4}{7}\right)^{\text{th}}$ of the Barbies were purchased before the age of 10. If 90 dolls in Ghazal's collection are Barbies that were purchased at the age of 10 or later, how many dolls in her collection are non-Barbie dolls?

(A) 70
(B) 90
(C) 140
(D) 154
(E) 192

79. The ratio of the ages of John and Suzy is 5 : 6. Which of the following can be the ratio of their ages after 10 years?

(A) 2 : 3
(B) 13 : 20
(C) 11 : 15
(D) 4 : 5
(E) 9 : 10

80. A company assembles two kinds of phones: a feature phone and a smartphone. Of the phones produced by the company last year, two-fifths were feature phones, and the rest were smartphones. If it takes $\frac{8}{5}$ times as many hours to produce a smartphone as it does to produce a feature phone, then the number of hours it took to produce the smartphones last year was what fraction of the total number of hours it took to produce all the phones?

(A) $\frac{8}{31}$

(B) $\frac{11}{31}$

(C) $\frac{12}{17}$

(D) $\frac{13}{34}$

(E) $\frac{15}{34}$

81. At a certain garment shop, the ratio of the number of shirts to the number of trousers is 4 to 5, and the ratio of the number of jackets to the number of shirts is 3 to 8. If the ratio of the number of sweaters to the number of trousers is 6 to 5, what is the ratio of the number of jackets to the number of sweaters?

(A) 9 to 25

(B) 1 to 3

(C) 1 to 4

(D) 3 to 5

(E) 6 to 5

82. At a church prayer, $\left(\frac{3}{5}\right)^{\text{th}}$ of the members were males. $\left(\frac{3}{5}\right)^{\text{th}}$ of the male members and $\left(\frac{7}{10}\right)^{\text{th}}$ of the female members attended the prayer. Of the members who did not attend the prayer, what fraction are male members who did not attend the prayer?

(A) $\frac{1}{4}$

(B) $\frac{3}{7}$

(C) $\frac{2}{3}$

(D) $\frac{9}{10}$

(E) $\frac{6}{19}$

83. John, Suzy, and David together donated a total of \$100 to a charity. If John paid $\left(\frac{5}{3}\right)^{\text{th}}$ of what David donated, Suzy donated \$20, and David donated the rest, what fraction of the total amount did David donate?

(A) $\frac{1}{5}$

(B) $\frac{1}{6}$

(C) $\frac{2}{7}$

(D) $\frac{3}{10}$

(E) $\frac{4}{11}$

84. A merchant sold a total of X shirts and trousers. If the number of trousers is $\left(\frac{1}{5}\right)^{\text{th}}$ the number of shirts, and $\left(\frac{1}{5}\right)^{\text{th}}$ of the shirts are cotton shirts, how many cotton shirts, in terms of X, were sold by the merchant?

(A) $\frac{2X}{7}$

(B) $\frac{X}{4}$

(C) $\frac{4X}{15}$

(D) $\frac{6X}{25}$

(E) $\frac{X}{6}$

85. A rod that weighs 20 pounds is cut into two pieces such that one of the pieces weighs 16 pounds and measures 36 feet. If the weight of the rod is directly proportional to the square of its length, how long is the other piece of rod, in feet?

(A) 9

(B) 12

(C) 18

(D) 24

(E) 27

86. The ratio of John's coins to Suzy's coins is 3 : 4. If the number of John's coins exceeds $\frac{2}{7}$ of the total number of coins by 25, how many coins Suzy has?

(A) 50

(B) 100

(C) 120

(D) 150

(E) 180

87. The total cost of an item is formed out of four costs: Material cost, Labour cost, Factory overhead cost, and Office overhead cost. If Material cost and Labour cost constitute $\frac{3}{7}$ part of the total cost, Labour cost and Factory overhead cost constitute $\frac{1}{2}$ part of the total cost, Factory overhead cost and Office overhead cost constitute $\frac{4}{7}$ part of the total cost, and Material cost and Office overhead cost constitute $\frac{1}{2}$ part of the total cost, which of the four costs is the highest among all?

(A) Material cost

(B) Labour cost

(C) Factory overhead cost

(D) Office overhead cost

(E) Factory overhead cost or Office overhead cost

88. An entrance test consists of 20 questions. Each question after the first is worth 2 points more than the preceding question. If the total questions are worth a total of 400 points, how many points is the fourth question worth?

(A) 5
(B) 7
(C) 11
(D) 19
(E) 38

89. In a science college, 80 more than $\left(\frac{1}{3}\right)^{\text{rd}}$ of all the students took a science course and $\left(\frac{1}{3}\right)^{\text{th}}$ of those who took the science course took chemistry. If $\left(\frac{1}{6}\right)^{\text{th}}$ of all the students in the school took chemistry, how many students are in the school?

(A) 200
(B) 240
(C) 480
(D) 600
(E) 720

90. In an office having 50 employees, $\left(\frac{1}{4}\right)^{\text{th}}$ of the males and $\left(\frac{1}{5}\right)^{\text{th}}$ of the females eat company breakfast. What is the greatest possible number of employees in the office that eat company breakfast?

(A) 6
(B) 8
(C) 10
(D) 12
(E) 25

91. In a cookery class, $\left(\frac{1}{8}\right)^{\text{th}}$ of the number of females is equal to $\left(\frac{1}{12}\right)^{\text{th}}$ of the total number of students in the class. What is the ratio of the number of males to the number of females in the class?

(A) 1 : 5
(B) 1 : 4
(C) 1 : 2
(D) 3 : 4
(E) 2 : 1

92. In a class, there are 40 students. In a test, the average (arithmetic mean) score of the girls is 30, that of the boys is 40, and that of the class is 32. If the score of a boy was incorrectly computed at 30 for 40, what is the correct average (arithmetic mean) score of the class?

(A) 31
(B) 31.50
(C) 32.25

(D) 41

(E) 42

93. How many liters of Chemical A must be added to a 120-liter solution that is 25 percent Chemical A in order to produce a solution that is 40 percent Chemical A?

(A) 12

(B) 15

(C) 20

(D) 24

(E) 30

94.

Month	Number of chickens
1	144
2	c
3	256

Information about the number of chickens hatched in a poultry farm is given in the table above. If the number of chickens in the poultry farm in any month increased by the same fraction during each of the two periods of the successive months, how many chickens were there in the second month?

(A) 192

(B) 200

(C) 210

(D) 220

(E) 240

2.6 Speed, Time & Distance

95. A trip of 900 miles would have taken 1 hour less if the average speed for the trip had been greater by 10 miles per hour. What was the average speed for the trip?

(A) 40 miles per hour
(B) 45 miles per hour
(C) 60 miles per hour
(D) 75 miles per hour
(E) 90 miles per hour

96. A truck traveled 336 miles per full tank of diesel on the national highway and 224 miles per full tank of diesel on the state highway. If the truck traveled 4 fewer miles per gallon on the state highway than on the national highway, how many miles per gallon did the truck travel on the state highway?

(A) 6
(B) 8
(C) 10
(D) 12
(E) 15

97. A bike traveling at a certain constant speed takes 5 minutes longer to travel 10 miles than it would take at 60 miles per hour. At what speed, in miles per hour, is the bike traveling?

(A) 36
(B) 40
(C) 42
(D) 48
(E) 50

98. A biker increased his average speed by 10 miles per hour in each successive 10-minute interval after the first interval. If in the first 10-minute interval, his average speed was 30 miles per hour, how many miles did he travel in the fourth 10-minute interval?

(A) 4
(B) 5
(C) 8
(D) 10
(E) 15

99. An aircraft flew 600 miles to a town at an average speed of 500 miles per hour with the wind and made the trip back following the same route at an average speed of 400 miles per hour against the wind. Which of the following is aircraft's approximate average speed, in miles per hour, for the trip?

(A) 420
(B) 444

(C) 450
(D) 467
(E) 483

100. A truck completed half of a 800-mile trip at an average speed of 40 miles per hour. At what approximate average speed, in miles per hour, should the truck complete the remaining miles to achieve an average speed of 50 miles per hour for the entire 800-mile trip? Assume that he truck completed its 800-mile trip without stoppage.

(A) 52
(B) 55
(C) 60
(D) 67
(E) 70

101. A marathoner ran for two days. On the second day, he ran at an average speed of 3 miles per hour, faster than the average speed of the first day. If during the two days he ran a total of 36 miles and did a total of 8 hours running, which of the following could be his average speed, in miles per hour, on the first day?

(A) 0.25
(B) 0.50
(C) 1.00
(D) 1.50
(E) 2.00

102. Two trains traveling toward each other on parallel tracks at constant rates of 50 miles per hour and 60 miles per hour are 285 miles apart. How far apart will they be 2 hours before their engine meet?

(A) 110
(B) 120
(C) 150
(D) 200
(E) 220

103. If the speed limit along an 10-mile section of rail track is reduced from 50 miles per hour to 40 miles per hour. Approximately how many minutes more will it take a train to travel along this section with the new speed limit than it would have taken at the old speed limit?

(A) 3
(B) 5
(C) 8
(D) 10
(E) 12

104. Trains A and B traveled the same 100-mile route. If Train A took 4 hours and Train B traveled at an average speed 25 percent more than the average speed of Train A, how many hours did it take Train B to travel the route?

(A) $1\frac{2}{3}$

(B) 2

(C) $3\frac{1}{5}$

(D) $3\frac{3}{5}$

(E) 4

105. Jeff drives three times farther in 36 minutes than Amy drives in 30 minutes. If Jeff drives at a speed of 40 miles per hour, at what speed, in miles per hour, does Amy drive?

(A) 6

(B) 9

(C) 16

(D) 24

(E) 32

106. A bus left bus depot A at 7 a.m. and reached bus depot B at 12 p.m. Another bus left bus depot B at 8 a.m. and reached bus depot A at 11 a.m. At what time did the two buses pass each other?

(A) 9:00 am

(B) 9:15 am

(C) 9:30 am

(D) 9:40 am

(E) 10:00 am

2.7 Time & Work

107. Photocopier machine A makes 1,500 copies per hour. Working 12 hours each day, photocopier machine B, twice as efficient as A, will make how many copies in 20 days?

(A) 400,000

(B) 500,000

(C) 540,000

(D) 660,000

(E) 720,000

108. A water pump began filling an empty swimming pool and ran at a constant rate till the pool was full. At some point, the pool was half full, and $2\frac{1}{3}$ hours later, it was $\frac{5}{6}$ full. How many hours did it take the pump to fill the empty pool completely?

(A) 4

(B) $5\frac{1}{3}$

(C) 7

(D) $7\frac{1}{5}$

(E) $8\frac{1}{3}$

109. Two pumps, each working alone, can fill an empty pool in 10 hours and 15 hours, respectively. The first pump initially started alone for h hours, after which the second pump was also started. If it took a total of 7 hours for the pool to be filled completely, what is the value of h?

(A) 2.00

(B) 2.50

(C) 3.00

(D) 3.30

(E) 4.00

110. An empty swimming pool with a capacity of 5,760 gallons is being filled at a rate of 12 gallons per minute by a pipe. An emptying pipe that has the capacity to empty $\left(\frac{3}{4}\right)^{\text{th}}$ of the pool in 9 hours is also in operation. If the pool is already half-filled and if both pipes are in operation, how many hours would it take to fill the pool to its full capacity?

(A) 6

(B) 12

(C) 24

(D) 36

(E) 72

111. Lathe machine A manufactures metal parts three times as fast as lathe machine B. Lathe machine B manufactures 300 X-type bearings in 60 days. If each machine manufactures bearings at their constant rate, how many Y-type bearings does lathe machine A manufacturer in 10 days if each Y-type bearing takes 2.5 times the time taken to manufacture each X-type bearing?

(A) 40
(B) 50
(C) 54
(D) 60
(E) 64

112. Photocopier A, working alone at its constant rate, makes 1,200 copies in 3 hours. Photocopier B, working alone at its constant rate, makes 1,200 copies in 2 hours. Photocopier C, working alone at its constant rate, makes 1,200 copies in 6 hours. How many hours will it take photocopiers A, B, and C, working together at their respective constant rates, to make 3,600 copies?

(A) 2.00
(B) 2.25
(C) 2.50
(D) 3.00
(E) 3.50

113. Five men can consume food costing \$150 on a 4-day expedition trip. If a woman consumes three-fourths of the amount of food consumed by a man, what would be the cost of food consumed by 4 men and 2 women during an 8-day expedition trip?

(A) \$300
(B) \$330
(C) \$360
(D) \$390
(E) \$400

114. Working alone, Mark and Kate take 12 hours more and 27 hours more, respectively, to complete a certain project than working together would take them to complete the same project. Working together, how many hours do Mark and Kate take to complete the project?

(A) 12
(B) 16
(C) 18
(D) 24
(E) 39

2.8 Computational

115. A chemical evaporates out of a beaker at the rate of x liters for every y minutes. If the chemical costs 25 dollars per liter, what is the cost, in dollars, of the amount of the chemical that will evaporate in z minutes?

(A) $\frac{25x}{yz}$

(B) $\frac{xz}{25q}$

(C) $\frac{25y}{xz}$

(D) $\frac{25xz}{y}$

(E) $\frac{25yz}{x}$

116. In company X, the total cost of producing pens is governed by a linear function. If the total cost of producing 25,000 pens is \$37,500 and the total cost of producing 35,000 pens is \$47,500, what is the total cost of producing 50,000 pens?

(A) \$57,500

(B) \$60,000

(C) \$62,500

(D) \$67,857

(E) \$75,900

117. If Suzy had thrice the amount of money that she currently has, she would have exactly the money needed to purchase four pencils, each costing \$1.35 per piece, and two erasers, each costing \$0.30 per piece. How much money does Suzy have?

(A) \$1.50

(B) \$2.00

(C) \$2.25

(D) \$2.50

(E) \$2.75

118. The population of a certain country increases at the rate of 30,000 people every month. The population of the country in 2012 was 360 million. In which year would the population of the country be 378 million?

(A) 2060

(B) 2061

(C) 2062

(D) 2063

(E) 2064

119. An Ice cream parlor buys milk-cream cartons, each containing $2\frac{1}{2}$ cups of milk-cream. If the restaurant uses $\frac{1}{2}$ cup of milk-cream in each serving of its ice cream, what is the least number of cartons needed to prepare 98 servings of the ice cream?

(A) 9
(B) 19
(C) 20
(D) 21
(E) 24

120. A few coins are put into 7 boxes, such that each box contains at least two coins. At most 3 boxes can contain the same number of coins, and the remaining boxes cannot contain an equal number of coins. What is the minimum possible number of coins in the 7 boxes?

(A) 18
(B) 20
(C) 24
(D) 27
(E) 30

121. A volcanic lava laterally moves at a rate of $15/4$ feet per hour. How many days does it take the lava to move $3/2$ miles? (1 mile = 5,280 feet)

(A) 48
(B) 60
(C) 72
(D) 80
(E) 88

122. At a metal rolling factory, if an iron bar of square cross-section with an area of 4 square feet is moving continuously through a belt conveyor at a constant speed of 360 feet per hour, how many seconds does it take for a volume of 8.4 cubic feet of the iron bar to move through the conveyor?

(A) 21
(B) 22
(C) 24
(D) 27
(E) 30

123. At a factory, each worker is remunerated according to a salary grade G, from 1 to 7. Each worker receives a monthly wage W, in dollars, determined by the formula $W = 1,140 + 45(G - 1)$. How many more dollars per month does a worker with a salary grade of 7 receive than a worker with a salary grade of 1?

(A) \$135
(B) \$270
(C) \$405
(D) \$540
(E) \$600

124. At a garage sale, a few radios and DVD players were sold. The prices of all the items were different. Of the radios, the price of a certain radio sold was the 15th highest price as well as the 20th lowest price. Of all items, the price of a certain DVD player sold was the 29th highest price as well as the 37th lowest price. How many DVD players were sold at the garage sale?

(A) 30

(B) 31

(C) 32

(D) 64

(E) 65

125. A salesman is paid \$25 per order as commission for the first 150 orders and \$12.50 as commission for each additional order. If he received a total of \$5,000 as commission, how many orders did he make?

(A) 100

(B) 150

(C) 200

(D) 250

(E) 300

126. An overseas businessman purchased a total of \$2,000 worth of traveler's checks in \$20 and \$50 denominations. During the trip, he cashed only 10 checks and lost all the remaining checks. If the number of \$20 checks cashed were 2 more or 2 less than the number of \$50 checks cashed, what is the minimum possible amount of money he lost?

(A) \$1,200

(B) \$1,440

(C) \$1,500

(D) \$1,620

(E) \$1,680

127. A pet shop sells three pack-sizes of dog food from brand X. The 5-kg pack costs \$16, the 10-kg pack costs \$26, and the 25-kg pack costs \$55. If a customer wants to buy a minimum of 40 kg of brand X dog food, what is the minimum price he will have to pay?

(A) \$85

(B) \$90

(C) \$97

(D) \$107

(E) \$110

2.9 Interest

128. If a sum of money invested under simple interest, amounts to \$3,200 in 4 years and \$3,800 in 6 years, what is the rate at which the sum of money was invested?

(A) 10%

(B) 12%

(C) 15%

(D) 20%

(E) 24%

129. For an equal sum of money, each invested for 2 years, the difference between compound interest, compound annually, and simple interest, at the same rate of interest, is \$63. If the simple interest on the sum after 2 years is \$600, at what rate of interest were the sums of money invested?

(A) 25%

(B) 24%

(C) 22%

(D) 21%

(E) 10%

130. Suzy borrows two equal sums of money under simple interest at 10% and 8% rates of interest. She finds that if she repays the former sum one year before the latter, she would have to pay the same amount for each borrowing. After how many years did she pay the first sum of money?

(A) 2.5

(B) 3

(C) 3.5

(D) 4

(E) 5

131. A sum of \$100,000 was invested in two deposits: one part at simple interest rates of 3 percent and another part at simple interest rates of 4 percent. If the total interest on the two sums was \$3,600 at the end of one year, what fractional part of the \$100,000 was invested at 4 percent?

(A) $\frac{5}{8}$

(B) $\frac{1}{5}$

(C) $\frac{2}{5}$

(D) $\frac{3}{5}$

(E) $\frac{3}{7}$

132. A sum of money is invested at simple interest, partly at 4% and remaining at 7% annual rates of interest. After two years, the total interest obtained was \$2,100. If the total investment is \$18,000, what was the sum of money invested at a 4% annual rate of interest?

(A) \$5,500
(B) \$6,000
(C) \$7,000
(D) \$10,500
(E) \$11,000

133. A man invested two equal sums of money in two banks at simple interest, one offering an annual rate of interest of 10% and the other offering an annual rate of interest of 20%. If the difference between the interests earned after 2 years is between \$120 and \$140, exclusive, which of the following could be the difference between the amounts earned for the same sums of money, invested at the same rates of interest as above, but at compound interest?

(A) \$130
(B) \$135
(C) \$137
(D) \$154
(E) \$162

134. At the start of an experiment, a certain population consisted of x organisms. At the end of each month, the population size increased by twice as much as size at the beginning of that month. If the total population at the end of five months is greater than 1,000, what is the minimum possible value of x?

(A) 2
(B) 3
(C) 4
(D) 5
(E) 6

135. A sum of money is borrowed at a 12% per annum interest rate for one year. The interest is calculated after the end of every two-month period and is added to the amount accrued after a period. The amount payable after the end of the year is how many times the sum borrowed?

(A) 1.12
(B) $(1.12)^5$
(C) $(1.02)^6$
(D) $(1.02)^5$
(E) $(1.2)^6$

136. Mary deposited a sum of x dollars into an account that earned 4% annual interest compounded annually. One year later, she deposited an additional x dollars in the account. Consider that there were no other transactions, and if the account showed y dollars at the end of the two years, which of the following expresses x in terms of y?

(A) $\frac{y}{2.04}$
(B) $\frac{y}{1.12}$
(C) $\frac{y}{2.2}$

(D) $\frac{y}{1.04 \times 2.04}$

(E) $\frac{y}{(1.04)^2 \times 2.04}$

137. George invested a certain sum of money in compound interest payable at a certain rate of interest. By the end of the 5th year, the interest on the investment was \$4,800, and by the end of the 6th year, the interest on the investment was \$5,520. What was the rate of interest at which George invested the sum of money?

(A) 10.0%

(B) 12.0%

(C) 12.5%

(D) 15.0%

(E) 20.0%

2.10 Functions

138. If the function f is defined by $f(p) = p^2 + \frac{1}{p^2}$ for all non-zero numbers p, then $\left(f\left(-\frac{1}{\sqrt{p}}\right)\right)^2 =$

(A) $f(p) + 2$

(B) $\frac{2}{f(p^2)}$

(C) $\left(\frac{1}{f(\sqrt{p})}\right)^2$

(D) $1 - (f(\sqrt{p}))^2$

(E) $f(p) - 2$

139. The function f is defined by $f(x) = -\frac{1}{x}$ for all non-zero numbers x. If $f(a) = -\frac{1}{2}$ and $f(ab) = \frac{1}{6}$, then $b =$

(A) 3

(B) $\frac{1}{3}$

(C) $-\frac{1}{3}$

(D) -3

(E) -12

140. The function f is defined by $f(x) = \sqrt{x} - 20$ for all positive numbers x. If $p = f(q)$ for some positive numbers p and q, what is q in terms of p?

(A) $(p + 20)^2$

(B) $\sqrt{p + 20}$

(C) $(\sqrt{p} + 20)^2$

(D) $\sqrt{p^2 + 20}$

(E) $(p^2 + 20)^2$

141. The function f is defined for positive three-digit integer T by $f(T) = 2^a 3^b 5^c$, where a, b and c are the hundreds, tens and units digits of T, respectively. If K and R are three-digit positive integers such that $f(K) = 18f(R)$, then $K - R =$

(A) 65

(B) 70

(C) 80

(D) 100

(E) 120

142. For which of the following functions f, is $f(x) = f(1-x)$ for all x?

(A) $f(x) = 1 + x$

(B) $f(x) = 1 + x^2$

(C) $f(x) = x^2 - (1-x)^2$

(D) $f(x) = x^2(1-x)^2$

(E) $f(x^2) = \frac{x}{1-x}$

143. If $f(x) = \frac{1}{x}$ and $g(x) = \frac{x}{x^2+1}$, for all $x > 0$, what is the minimum value of $f(g(x))$?

(A) 0

(B) $\frac{1}{2}$

(C) 1

(D) $\frac{3}{2}$

(E) 2

144. If $f(x) = \frac{10x}{1-x}$, for what value of x does $f(x) = \frac{1}{2}f(3)$?

(A) 4

(B) 2

(C) 1

(D) -3

(E) -5

145. If $3f(x) + 2f(-x) = 5x - 10$, what is the value of $f(1)$?

(A) 0

(B) 1

(C) 2

(D) 3

(E) 4

146. As per an estimate, the depth $D(t)$, in centimeters, of the water in a tank at t hours past 12:00 a.m. is given by $D(t) = -10(t-7)^2 + 100$, for $0 \le t \le 12$. At what time does the depth of the water in the tank becomes the maximum?

(A) 5:30 a.m.

(B) 7:00 a.m.

(C) 7:30 a.m.

(D) 8:00 a.m.

(E) 9:00 a.m.

2.11 Permutation & Combination & Probability

147. $C_q^p = \frac{p!}{(p-q)! \times q!}$ for positive integers p and q, $p \geq q$. If $C_3^5 = C_r^5$, which of the following could be the value of r?

(A) 0
(B) 1
(C) 2
(D) 4
(E) 5

148. A color code is defined as a sequence of three dots arranged in a row. Each dot is colored either "red" or "black." How many distinct codes can be formed?

(A) 4
(B) 5
(C) 6
(D) 8
(E) 10

149. A daily store stocks two sizes of mugs, each in four colors: black, green, yellow, and red. The store packs the mugs in packages that contain either three mugs of the same size and the same color or three mugs of the same size and three different colors. If the order of the arrangement of mugs of three different colors is disregarded, how many different packings of the types described above are possible?

(A) 4
(B) 10
(C) 16
(D) 20
(E) 30

150. A pizza-seller offers six kinds of toppings and two kinds of breads for his pizzas. If each pizza contains at least two kinds of toppings but not all kinds of toppings and only one kind of bread, how many different pizzas could the pizza-seller offer?

(A) 56
(B) 58
(C) 84
(D) 100
(E) 112

151. A botanist designates each plant with a one-, two-, or three-letter code, where each letter is one of the 26 letters of the alphabet. If the letters may be repeated and if the same letters used in a different order convey a different code, how many different plants can the botanist uniquely designate with these codes?

(A) 2,951

(B) 9,125
(C) 16,600
(D) 17,576
(E) 18,278

152. A college student can select one out of eight optional subjects from group one and two out of ten optional subjects from group two. If no subject is common in both groups, how many different sets of three subjects are there to select?

(A) 53
(B) 120
(C) 190
(D) 360
(E) 408

153. A company has to assign distinct four-digit code numbers to its employees. Each code number was formed from the digits 1 to 9, and no digit appears more than once in any one code. How many employees can be assigned codes?

(A) 30
(B) 2,400
(C) 3,024
(D) 5,040
(E) 6,491

154. A company plans to assign identification numbers to its employees. Each number is to consist of four digits from 0 to 9, inclusive, except that the first digit cannot be 0. If any digit can be repeated any number of times in a particular code, how many different identification numbers are possible that are odd numbers?

(A) 2,520
(B) 2,268
(C) 3,240
(D) 4,500
(E) 9,000

155. A fast-food company plans to build four new restaurants. There are six sites (A, B, C, D, E, and F) that satisfy the company's criteria for the location of the restaurants. In how many ways can the company select four sites if both sites A and B cannot be selected together?

(A) 4
(B) 5
(C) 6
(D) 9
(E) 15

156. Imran has four Math, five Physics, and six Chemistry books. He has to choose four out of the 15 books such that the selection has at least one book of each subject. In how many ways it is possible?

(A) 600
(B) 720
(C) 760
(D) 800
(E) 960

157. A botanist plans to code the 15 plants used in an experiment with unique codes that consist of either a single letter or a pair of distinct letters. If the codes must be in alphabetic order, i.e., a code CD is valid but a code DC is not, what is the least number of letters that can be used to code them?

(A) 3
(B) 4
(C) 5
(D) 7
(E) 15

158. Classes A, B, and C have 30 students each, and class D has 20 students. A team is to be formed by selecting one student from A, B, and C and two students from D. How many teams are possible?

(A) 1,540,000
(B) 2,200,000
(C) 2,400,000
(D) 3,600,000
(E) 5,130,000

159. A stock broker recommends a portfolio of 2 Information Technology stocks, 4 Retail stocks, and 2 e-commerce stocks. If the broker can choose from 4 Information Technology stocks, 5 Retail stocks, and 3 e-commerce stocks, how many different portfolios of 8 stocks are possible?

(A) 9
(B) 24
(C) 60
(D) 90
(E) 120

160. In a conference of 3 delegates from each of 8 different companies, each delegate shook hands with every person other than those from his or her own organization. How many handshakes took place at the conference?

(A) 48
(B) 96
(C) 252
(D) 270
(E) 504

161. The digits 0 to 9 are used to form three-digit codes; however, there are a few conditions: the first digit cannot be 0 or 9, the second digit must be 0 or 9, and the second and third digits cannot both be '9' in the same code. If the digits may be repeated in the same code, how many different codes are possible?

(A) 152

(B) 156

(C) 160

(D) 729

(E) 1,000

162. A pot contains 15 green and 15 yellow marbles. If two marbles are to be picked from this pot at random, what is the probability that both marbles will be yellow?

(A) $\frac{1}{5}$

(B) $\frac{7}{29}$

(C) $\frac{7}{30}$

(D) $\frac{8}{29}$

(E) $\frac{23}{30}$

163. A box contains 12 balls; of these, seven are red and five are green. If three balls are to be selected at random from the box, what is the probability that two of the balls will be red and one will be green?

(A) $\frac{7}{44}$

(B) $\frac{7}{22}$

(C) $\frac{51}{100}$

(D) $\frac{21}{44}$

(E) $\frac{7}{9}$

164. A badminton club has 21 members. What is the ratio of the number of 6-member committees the number of 5-member committees that can be formed from the members of the club?

(A) 6 to 5

(B) 15 to 1

(C) 8 to 3

(D) 17 to 6

(E) 16 to 5

165. A courier company can assign its employees to its offices in such a way that more than one office can be assigned to an employee. In how many ways can the company assign four employees to two offices?

(A) 6

(B) 8

(C) 10

(D) 12

(E) 16

166. A transport company employs five male officers and three female officers. If a core group is to be created that is made up of three male and two female officers, how many different core groups are possible?

(A) 10

(B) 16

(C) 24

(D) 30

(E) 60

167. If the probability that Stock X will increase in value during the next week is 0.40 and the probability that Stock Y will increase in value during the next week is 0.60, what is the probability that exactly one of the two stocks will increase in value during the next week? Note that the price fluctuations of one stock do not affect the price fluctuations of the other.

(A) 0.48

(B) 0.50

(C) 0.52

(D) 0.56

(E) 0.58

168. An unbiased coin has an equal probability of getting a head or a tail. What is the probability that the coin will land heads at least once when it is tossed twice?

(A) $\frac{1}{5}$

(B) $\frac{1}{4}$

(C) $\frac{1}{3}$

(D) $\frac{2}{3}$

(E) $\frac{3}{4}$

169. A quiz consists of X questions, each of which is to be answered either "Yes" or "No." The participant must answer all questions correctly to win the quiz. What is the least value of X for which the probability is less than $\frac{1}{500}$ such that a participant who randomly guesses the answer to each question will be a winner?

(A) 8

(B) 9

(C) 10

(D) 200

(E) 500

170. A box contains 20 balls: 10 white and 10 black. Five balls are to be drawn at random. If the first three drawn balls are black and not replaced, what is the probability that the next two drawn balls will also be black?

(A) $\frac{21}{136}$

(B) $\frac{4}{17}$

(C) $\frac{1}{3}$

(D) $\frac{4}{7}$

(E) $\frac{2}{5}$

171. A box contains 16 balls, of which 4 are white, 3 are blue, and the rest are yellow. If two balls are to be selected at random from the box, one at a time, without being replaced, what is the probability that one ball selected will be white and the other ball selected will be blue?

(A) $\frac{5}{64}$

(B) $\frac{1}{16}$

(C) $\frac{1}{10}$

(D) $\frac{1}{5}$

(E) $\frac{1}{6}$

172. A batch of eight refrigerators contains two single-door refrigerators and six double-door refrigerators. If two refrigerators are to be chosen at random from this batch, what is the probability that at least one of the two refrigerators chosen will be a single-door?

(A) $\frac{1}{4}$

(B) $\frac{1}{3}$

(C) $\frac{7}{15}$

(D) $\frac{13}{28}$

(E) $\frac{1}{2}$

173. In a jar, 9 balls are white, and the rest are red. If two balls are to be chosen at random from the jar without replacement, the probability that the balls chosen will both be white is $\frac{6}{11}$. What is the number of balls in the jar?

(A) 10

(B) 11

(C) 12

(D) 13

(E) 15

174. A pyramid of 12 playing cards is in such a state that if any individual card falls, the pyramid collapses. If, for each individual card, the probability of falling within 1 minute is 0.05, what is the probability that the pyramid will collapse within 1 minute?

(A) 0.05

(B) 0.05^{12}

(C) $1 - 0.95^{12}$

(D) 0.95^{12}

(E) $1 - 0.05^{12}$

175. IIn a pack of a dozen candies, four are orange-flavored. If a kid randomly picks two candies from the pack, what is the probability that the kid has no orange-flavored candy?

(A) $\frac{1}{7}$

(B) $\frac{2}{11}$

(C) $\frac{14}{33}$

(D) $\frac{7}{33}$

(E) $\frac{8}{33}$

176. On the morning of day 1, Suzy began her tracking tour. She plans to return home at the end of the first day on which it rains. If for the first three days of the tour, the probability of rain on each day is 0.25, what is the probability that Suzy will return home at the end of day 3?

(A) $\frac{1}{8}$

(B) $\frac{9}{64}$

(C) $\frac{27}{64}$

(D) $\frac{37}{64}$

(E) $\frac{1}{64}$

2.12 Sets

177. A marketing class at a college has a total strength of 30. It formed three groups: G1, G2, and G3, which have 10, 10, and 6 students, respectively. If no student in G1 is in either G2 or G3, what is the greatest possible number of students who are in neither of the groups?

(A) 4
(B) 7
(C) 8
(D) 10
(E) 14

178. In a batch of 1,000 dresses, 1/4 of the dresses are traditional and the rest are contemporary. Half the dresses are for males, and half are for females. If of all dresses, 100 dresses are traditional and are for males, how many of the dresses are contemporary and are for females?

(A) 150
(B) 250
(C) 300
(D) 350
(E) 400

179. According to a report, 7% of students did not use computers to play games, 11% did not use computers to write reports, and 95% use computers for at least one of the purposes. What percent of the students, according to the report, used computers for both purposes—playing games and writing reports?

(A) 13%
(B) 56%
(C) 77%
(D) 87%
(E) 91%

180. In a company survey, 600 employees were each asked whether they take cola or health drinks. As per the survey, 70 percent of the employees take cola, 45 percent take health drinks, and 25 percent take both cola and health drinks. How many employees surveyed take neither cola nor health drinks?

(A) 50
(B) 60
(C) 70
(D) 75
(E) 80

181. In Milton School, the number of students who play badminton is three times the number of students who play tennis. The number of students who play both badminton and tennis is three times the number of students who play only tennis. If 60 students play both sports, how many students play only badminton?

(A) 100

(B) 150

(C) 160

(D) 180

(E) 120

2.13 Statistics & Data Interpretation

182. 15, 20, 25, x: (not in order)

Which of the following could be the median of the four integers listed above?

I. 18
II. 22
III. 23

(A) I only
(B) II only
(C) I and II only
(D) II and III only
(E) All of them

183. 44, 52, 56, 65, 73, 75, 77, 95, 96, 97

The list above shows the scores of 10 students obtained on a scheduled test. If the standard deviation of the 10 scores is 20.50 and the arithmetic mean is 73, how many scores are greater than one standard deviation above the mean of the 10 scores?

(A) None
(B) One
(C) Two
(D) Three
(E) Four

184. A set consists of 20 numbers. If n is a number in the list and is four times the average (arithmetic mean) of the numbers in the list, excluding itself, then n is what fraction of the sum of the 20 numbers in the list?

(A) $\frac{1}{20}$
(B) $\frac{4}{23}$
(C) $\frac{1}{5}$
(D) $\frac{1}{10}$
(E) $\frac{5}{11}$

185. If the average (arithmetic mean) of 3, 8, and w is greater than or equal to w and smaller than or equal to $3w$, how many integer values of w exist?

(A) Five
(B) Four
(C) Three
(D) Two
(E) One

186. If the average (arithmetic mean) of seven distinct positive integers is 14, what is the least possible value of the greatest of the seven numbers?

(A) 14
(B) 17
(C) 18
(D) 20
(E) 77

187. If the average (arithmetic mean) of x, y and 10 is equal to the average of x, y, 10 and 20, what is the sum of x and y?

(A) 40
(B) 50
(C) 55
(D) 60
(E) 65

188. A set of 13 different integers has a median of 20 and a range of 20. What is the greatest possible value of the integer in the set?

(A) 23
(B) 27
(C) 30
(D) 34
(E) 40

189. The mean of the set of seven positive integers 1, 2, 3, 4, 5, 6, and x is $\frac{2x}{7}$. What is the value of x?

(A) 1
(B) 7
(C) 14
(D) 18
(E) 21

190. A company has a total of x employees, such that no two employees have the same annual salary. The annual salaries of the x employees are listed in increasing order, and the 22nd salary in the list is the median of their annual salaries. If the sum of the annual salaries of all the employees is \$860,000, what is the average (arithmetic mean) of the annual salaries of all the employees?

(A) \$19,500
(B) \$20,000
(C) \$25,000
(D) \$30,000
(E) \$32,500

191. The table below gives information about the electricity consumption of four appliances in a household. What is the average number of watts of electricity used per hour per appliance in the household?

Electricity usage in the household		
Appliance	**Number of hours in use**	**Number of watts of electricity used per hour**
Computer	4	105
Music system	2	90
Refrigerator	2	235
LED TV	2	150

(A) 76
(B) 105
(C) 137
(D) 187
(E) 303

192. The heights a large population have a distribution that is symmetric about the mean $\bar{x}$. If 68 percent of the distribution lies within one standard deviation d of the mean, what percent of the distribution is greater than $(\bar{x} - d)$?

(A) 16%
(B) 50%
(C) 68%
(D) 84%
(E) 95%

2.14 Linear Equations

193. A seller mistakenly reversed the digits of a customer's amount of change and returned an incorrect amount of change. If the customer received 63 cents more than he should have, which of the following could be the correct amount of change, in cents, he should have got?

(A) 89

(B) 36

(C) 38

(D) 37

(E) 29

194. A merchant sold screwdrivers for \$11 each and spanners for \$3 each. If a customer purchased a few screwdrivers and spanners for a total of \$109, what would be the total number of screwdrivers and spanners the customer purchased?

(A) 10

(B) 13

(C) 15

(D) 22

(E) 32

195. If $x + y + z = 2$, and $x + 2y + 3z = 6$ and $y \neq 0$, then what is the value of $\frac{x}{y}$?

(A) $-\frac{1}{2}$

(B) $-\frac{1}{3}$

(C) $-\frac{1}{6}$

(D) $\frac{1}{3}$

(E) $\frac{1}{2}$

196. A stationary shop sells a book for \$25 per piece and a notebook for \$15 per piece. In the previous month, it sold 2 more books than notebooks. If the total revenue from the sale of books and notebooks in the previous month was \$490, what was the total number of books and notebooks that the shop sold in the previous month?

(A) 18

(B) 20

(C) 24

(D) 30

(E) 33

2.15 Quadratic Equations & Polynomials

197. If $a \geq 0$ and $a = \sqrt{8ab - 16b^2}$, then in terms of b, $a =$

(A) $-4b$

(B) $\frac{b}{4}$

(C) b

(D) $4b$

(E) $4b^2$

198. What is the difference between the maximum and the minimum value of $\frac{x}{y}$ for which $(x-2)^2 = 9$ and $(y-3)^2 = 25$?

(A) $-\frac{15}{8}$

(B) $\frac{3}{4}$

(C) $\frac{9}{8}$

(D) $\frac{19}{8}$

(E) $\frac{25}{8}$

199. If x and y are positive integers and $2x + 3y + xy = 12$, what is the value of $(x + y)$?

(A) 2

(B) 4

(C) 5

(D) 6

(E) 8

200. A ball thrown up in air is at a height of h feet, t seconds after it was thrown, where $h = -3(t-10)^2 + 250$. What is the height of the ball once it reached its maximum height and then descended for 7 seconds?

(A) 96 feet

(B) 103 feet

(C) 164 feet

(D) 223 feet

(E) 250 feet

2.16 Inequalities

201. Suzy's college is 12 kilometers from her hostel. She travels 6 kilometers from the college to basketball practice, and from there, 4 kilometers for the computer class. If she is then D kilometers away from her hostel, what is the range of possible values for D?

(A) $1 \le D \le 5$

(B) $2 \le D \le 6$

(C) $2 \le D \le 10$

(D) $2 \le D \le 22$

(E) $4 \le D \le 24$

202. $2a + b = 12$, and $|b| \le 12$

How many ordered pairs (a, b) are solutions of the above system such that a and b both are integers?

(A) 9

(B) 10

(C) 11

(D) 12

(E) 13

203. If the cost of 15 pencils varies between \$3.60 and \$4.80 and the cost of 21 pens varies between \$33.30 and \$42.90, then the cost of 5 pencils and 7 pens varies between

(A) \$8.20 and \$12.20

(B) \$8.30 and \$10.20

(C) \$10.20 and \$16.30

(D) \$12.30 and \$15.90

(E) \$13.30 and \$16.60

204. Given that x is a negative number and $0 < y < 1$, which of the following is the greatest?

(A) x^2

(B) $(xy)^2$

(C) $\left(\frac{x}{y}\right)^2$

(D) $\frac{x^2}{y}$

(E) x^2y

205. David traveled from City A to City B in 5 hours, and his average speed was between 20 miles per hour and 30 miles per hour, while Mark traveled from City A to City B along the same route in 3 hours, and his average speed was between 40 miles per hour and 60 miles per hour. Which of the following could be the distance, in miles, from City A to City B?

(A) 105

(B) 135
(C) 155
(D) 160
(E) 165

2.17 Co-ordinate geometry

206. In the coordinate plane, a diameter of a circle has the end points $(-3,-6)$ and $(5,0)$. What is the area of the circle?

(A) 5π

(B) $10\sqrt{2}\pi$

(C) 25π

(D) 50π

(E) 100π

207. A straight line in the XY-plane has a slope of 3 and a Y-intercept of 4. On this line, what is the X-coordinate of the point whose Y-coordinate is 10?

(A) 2

(B) 4

(C) 6

(D) 7

(E) 7.5

208. In the XY-plane, a line l passes through the origin and has a slope 3. If points $(1,\ a)$ and $(b,\ 2)$ are on the line l, what is the value of $\frac{a}{b}$?

(A) 2

(B) 3

(C) $\frac{2}{3}$

(D) $\frac{2}{9}$

(E) $\frac{9}{2}$

209. In the XY-plane, the point $(3,2)$ is the center of a circle. The point $(-1,2)$ lies inside the circle, and the point $(3,-4)$ lies outside the circle. Which of the following could be the value of r?

(A) 5

(B) 4

(C) 3

(D) 2

(E) 1

210. In the XY-plane, the three vertices of a square are represented by points (a,b), $(a,-b)$, and $(-a,-b)$. If $a < 0$ and $b > 0$, which of the following points is in the same quadrant as the fourth vertex point of the square?

(A) $(-2,-6)$

(B) $(-2,6)$

(C) $(2,-6)$

(D) $(6, -2)$

(E) $(6, 2)$

211. In the XY-plane, the vertices of a triangle have coordinates $(0, 0)$, $(5, 5)$ and $(10, 0)$. What is the perimeter of the triangle?

(A) 12

(B) 13

(C) $5 + 10\sqrt{2}$

(D) $10 + 5\sqrt{2}$

(E) $10 + 10\sqrt{2}$

212. If the points $(a, 0)$, $(0, b)$ and $(1, 1)$ are collinear, what is the value of a in terms of b?

(A) $\frac{b-1}{b}$

(B) $\frac{b}{b+1}$

(C) $\frac{b}{b-1}$

(D) $\frac{b+1}{b}$

(E) $\frac{1}{b-1}$

213. In the XY-plane, what is the area of the triangle formed by the line $3y - 4x = 24$ and the X and Y axes?

(A) 6

(B) 14

(C) 24

(D) 36

(E) 48

Chapter 3

Answer key

(1) E	(25) C	(49) B	(73) B	(97) B
(2) C	(26) C	(50) A	(74) A	(98) D
(3) D	(27) C	(51) D	(75) C	(99) B
(4) D	(28) D	(52) C	(76) C	(100) D
(5) E	(29) B	(53) D	(77) C	(101) E
(6) C	(30) A	(54) C	(78) C	(102) E
(7) C	(31) D	(55) C	(79) E	(103) A
(8) A	(32) C	(56) E	(80) C	(104) C
(9) D	(33) D	(57) B	(81) C	(105) C
(10) A	(34) C	(58) D	(82) C	(106) C
(11) C	(35) D	(59) E	(83) D	(107) E
(12) E	(36) B	(60) A	(84) E	(108) C
(13) A	(37) D	(61) E	(85) C	(109) B
(14) C	(38) D	(62) C	(86) B	(110) B
(15) E	(39) B	(63) D	(87) E	(111) D
(16) D	(40) D	(64) C	(88) B	(112) D
(17) A	(41) A	(65) B	(89) C	(113) B
(18) D	(42) D	(66) D	(90) D	(114) C
(19) C	(43) B	(67) B	(91) C	(115) D
(20) C	(44) B	(68) C	(92) C	(116) C
(21) C	(45) B	(69) B	(93) E	(117) B
(22) A	(46) C	(70) B	(94) A	(118) C
(23) B	(47) B	(71) E	(95) E	(119) C
(24) D	(48) B	(72) D	(96) B	(120) C

(121) E

(122) A

(123) B

(124) B

(125) D

(126) D

(127) C

(128) C

(129) D

(130) D

(131) D

(132) C

(133) D

(134) D

(135) C

(136) D

(137) D

(138) A

(139) D

(140) A

(141) E

(142) D

(143) E

(144) D

(145) D

(146) B

(147) C

(148) D

(149) C

(150) E

(151) E

(152) D

(153) C

(154) D

(155) D

(156) B

(157) C

(158) E

(159) D

(160) C

(161) A

(162) B

(163) D

(164) C

(165) E

(166) D

(167) C

(168) E

(169) B

(170) A

(171) C

(172) D

(173) C

(174) C

(175) C

(176) B

(177) D

(178) D

(179) D

(180) B

(181) D

(182) C

(183) D

(184) B

(185) B

(186) B

(187) B

(188) D

(189) E

(190) B

(191) C

(192) D

(193) E

(194) C

(195) A

(196) C

(197) D

(198) E

(199) B

(200) B

(201) D

(202) E

(203) D

(204) C

(205) B

(206) C

(207) A

(208) E

(209) A

(210) E

(211) E

(212) C

(213) C

Chapter 4

Solutions

4.1 Number properties

1. Here, the given expression is in the format $a^2 - b^2 = (a + b)(a - b)$

$=> 99,996^2 - 4^2 = (99,996 + 4)(99,996 - 4)$

$=> 99,996^2 - 4^2 = (99,996 + 4)(99,992)$

$=> 100,000 \times (100,000 - 8) = 10^5 \times (10^5 - 8)$

The correct answer is Option E.

2. The highest exponent of 5 in $n!$ can be calculated by adding the quotients (integer parts) when n is successively divided by 5:

If $[x]$ denotes the integer part of x, we have

$$\left[\frac{n}{5}\right] + \left[\frac{\left[\frac{n}{5}\right]}{5}\right] + \cdots = 6$$

Assuming that only $\left[\frac{n}{5}\right]$ equals 6, we have $n = 5 \times 6 = 30$

However, for $n = 30$, the actual value of the exponent of 5 is:

- $\frac{30}{5} = 6$
- $\frac{6}{5} = 1$

The exponent is $6 + 1 = 7$, which is greater by $7 - 6 = 1$ than what was required.

Thus, we need to reduce 30 by '1' multiple of 5 i.e. $1 \times 5 = 5$.

Thus, the approximate value of $n = 30 - 5 = 25$.

For $n = 25$, the actual value of the exponent of 5 is:

- $\frac{25}{5} = 5$
- $\frac{5}{5} = 1$

The exponent is $5 + 1 = 6$, which is the exact exponent required.

However, since we need to find the largest value of the highest exponent of '7' in $n!$, we need to check if any higher value of n is possible.

We have obtained the highest exponent of '5' equals to 6 in 25!

However, we can increase n till a point before the next multiple of 5 is included.

Thus, we can increase n to 29! as in 30 includes a multiple of '5.'

Let us verify:

For $n = 29$, the actual value of the exponent of 5 is:

- $\frac{29}{5} = 5.8 \equiv 5$
- $\frac{5}{5} = 1$

Thus the exponent is $5 + 1 = 6$, which is the exact exponent required.

Thus, the maximum value of $n! = 29!$

Thus, for $n = 29$, the value of the highest exponent of 7 is:

- $\frac{29}{7} = 4.14 \equiv 4$

Thus the highest exponent of '7' in 29! is 4.

The correct answer is Option C.

Alternate approach:

Let us count the number of multiples of '5' in $n!$.

The multiples of '5' would be there in 5, 10, 15, 20, 25, 30, ...

We should stop counting till we get 6 multiples. Since 5, 10, 15, and 20 would give one each, and 25 would give two multiples, the total count of multiples $= 1 + 1 + 1 + 1 + 2 = 6 =$ the number of required multiples to get the greatest value of a.

We see that if $n < 30$, the value of $a = 6$, thus the greatest value of $n = 29$.

Now let's count the number of multiples of '7' in $n! = 29!$.

The multiples of '7' would be there in 7, 14, 21, and 28.

There are one multiple in each of 7, 14, 21, and 28: a total of four multiples.

3. We have $0 < a < 1$

Also, $b = a^2$ and $c = \sqrt{a}$

Multiplying both sides of the inequality, $a < 1$, by a (since a is positive, multiplying a will not change the sign of inequality):

$a \times a < 1 \times a$

$=> a^2 < a$

Thus, we have

$a > a^2 = b$

Again, since $a < 1$, taking square root on both sides, we have $\sqrt{a} < 1$

Multiplying both sides of the above inequality by $\sqrt{a}$, we have

$\sqrt{a} \times \sqrt{a} < \sqrt{a}$

$=> a < \sqrt{a} = c$

Thus, we have

$a^2 < a < \sqrt{a}$

$=> b < a < c$

The correct answer is Option D.

Alternate approach:

For any number less than 1, its squares, cubes & higher order numbers would be less than the number, and its square roots, cube roots & n^{th} roots would be greater than the number.

Since $a = 0.999 < 1$, this follows that $b < a < c$.

4. Given that,

$p = \frac{1}{150} \times \frac{1}{151} \times \cdots \times \frac{1}{250}$

$q = \left(\frac{1}{150} \times \frac{1}{151} \times \cdots \times \frac{1}{250}\right) \times \frac{1}{251} = \frac{p}{251}$

Thus, we have

$p^{-1} + q^{-1} = \frac{1}{p} + \frac{1}{q}$

$= \frac{1}{p} + \frac{251}{p}$

$=> \frac{252}{p}$

The correct answer is Option D.

5. The sum of all integers from 1 to n is given by $\frac{n(n+1)}{2}$.

Thus, the sum of all integers from 1 to 100 $= \frac{100 \times 101}{2} = 50 \times 101$.

And, the sum of all integers from 1 to 50 $= \frac{50 \times 51}{2} = 25 \times 51$.

Thus, the sum of all integers from 51 to 100 $= 50 \times 101 - 25 \times 51 = 25(202 - 51) = 25 \times 151 = 3,775$.

The correct answer is Option E.

6. $x = \frac{1}{51} + \frac{1}{52} + \cdots + \frac{1}{60}$

Let us try to find the range of values of x.

We know that x is the sum of 10 terms.

Thus, we can say:

$x < \frac{1}{51} + \frac{1}{51} + \ldots (10 \text{ times})$

$=> x < \frac{10}{51} = \frac{1}{5.1}$

$=> x < \frac{1}{5}$

Again, we have

$x > \frac{1}{60} + \frac{1}{60} + \ldots (10 \text{ terms})$

$=> x > \frac{10}{60} = \frac{1}{6}$

Thus, we have

$\frac{1}{6} < x < \frac{1}{5}$

$=> 6 > \frac{1}{x} > 5$

Similarly, we can deduce that $7 > \frac{1}{y} > 6$

Thus, $\frac{1}{y} > \frac{1}{x}$

The correct answer is Option C.

7. A number, here $4p25q$ is divisible by 4, if the number formed by the last two digits of the number is divisible by 4.

Thus, the number $5q$ is divisible by 4.

This is possible if $q = 2$ or 6 (since both 52 and 56 are divisible by 4).

A number, here number $4p25q$ is divisible by 9, if the sum of the digits of the number is divisible by 9.

The sum of the digits $= 4 + p + 2 + 5 + q = 11 + p + q$.

Thus, $(11 + p + q)$ should be divisible by 9.

If $q = 2$:

$11 + p + q$

$= 11 + p + 2$

$= 13 + p$.

Thus, $(13 + p)$ is divisible by 9 if $p = 5$ (since $13 + 5 = 18$, which is divisible by 9).

If $q = 6$:

$11 + p + q$

$= 11 + p + 6$

$= 17 + p$.

Thus, $(17 + p)$ is divisible by 9 if $p = 1$ (since $17 + 1 = 18$, which is divisible by 9).

Thus we have two possible situations:

$p = 5,\ q = 2 => \frac{p}{q} = \frac{5}{2}$

OR

$p = 1,\ q = 6 => \frac{p}{q} = \frac{1}{6}$

Thus, the minimum value of $\frac{p}{q} = \frac{1}{6}$.

The correct answer is Option C.

8. We know that if $a\%$ of $(a - 2b)$ when added to $b\%$ of b, the value obtained is 0.

$=> a\% \times (a - 2b) + b\% \times b = 0$

$$=> \frac{a}{100} \times (a - 2b) + \frac{b}{100} \times b = 0$$

$$=> \frac{a^2 - 2ab}{100} + \frac{b^2}{100} = 0$$

$$=> \frac{a^2 - 2ab + b^2}{100} = 0$$

$=> a^2 - 2ab + b^2 = 0$

$=> (a - b)^2 = 0$

$=> a - b = 0$

$=> a = b$

Thus, only statement I is correct.

The correct answer is Option A.

9. According to the problem:

If m is in the set, $(m^2 + 3)$ is also in the set.

However, it does NOT imply that if $(m^2 + 3)$ is in the set, then m must be in the set.

What it does imply is that: If $(m^2 + 3)$ is NOT in the set, m is NOT in the set.

Thus, if we have $m = -1$ as a member of the set, $(m^2 + 3) = 4$ is also a member of the set.

Thus, statement II is correct.

Proceeding in the same way:

Since $m = 4$ is a member of the set, then $(m^2 + 3) = 19$ is a member of the set.

Thus, statement III is also correct.

The correct answer is Option D.

10. We have

$t_2 = 5$

$t_3 = 2 \times t_2 - 1$
$= 2 \times 5 - 1$
$= 9$

$t_3 = 2 \times t_2 - 1$
$= 2 \times 9 - 1$
$= 17$

Thus, we see that the sequence follows the pattern:

$t_1, 5, 9, 17, ...$

We can calculate the value of t_1 too.

$t_2 = 2 \times t_1 - 1$
$5 = 2 \times t_1 - 1$
$t_1 = 3$

Again, we see that the sequence follows the pattern:

$3, 5, 9, 17, ...$

We can rewrite the above terms as following.

$(2^1 + 1), (2^2 + 1), (2^3 + 1), (2^4 + 1)...$

From the analogy, we can deduce that $t_{10} = 2^{10} + 1$ and $t_9 = 2^9 + 1$

Thus, $t_{10} - t_9 = (2^{10} + 1) - (2^9 + 1)$

$= 2^{10} + 1 - 2^9 - 1$

$= 2^9(2 - 1)$

$= 2^9$

The correct answer is Option A.

11. $660 = 2^2 \times 3 \times 5 \times 11$

We see that there is an extra '2' since the exponent of 2 is 2.

This extra '2' can be combined with the other factors to generate different values.

Also, keeping the two 2s separate, the other factors may be combined to generate different values of m, n, p and q.

Thus, the possible combination for the distinct values of m, n, p and q are:

m	n	p	q
3	$2 \times 2 = 4$	5	11
2	$2 \times 3 = 6$	5	11
2	3	5	$2 \times 11 = 22$
2	3	11	$2 \times 5 = 10$

Thus, there are four possible combinations.

The correct answer is Option C.

12. Looking at the expression, we can deduce that $\sqrt{x^{2/3} + y^{2/3} + 2(xy)^{1/3}}$ and $\sqrt{x^{2/3} + y^{2/3} - 2(xy)^{1/3}}$ are basically $(a + b)^2$ and $(a - b)^2$, respectively, where $a = x^{1/3}$ and $b = y^{1/3}$

So, $\sqrt{x^{2/3} + y^{2/3} + 2(xy)^{1/3}} = \sqrt{a^2 + b^2 + 2ab} = \sqrt{(a + b)^2} = a + b$

Similarly, $\sqrt{x^{2/3} + y^{2/3} - 2(xy)^{1/3}} = \sqrt{a^2 + b^2 - 2ab} = \sqrt{(a - b)^2} = b - a$; since $b > a$

Note that $x = 125$ and $y = 216$, or, $y > x$; thus, $y^{1/3} > x^{1/3}$. Or $b > a$

So, $\left(\sqrt{x^{2/3} + y^{2/3} + 2(xy)^{1/3}} + \sqrt{x^{2/3} + y^{2/3} - 2(xy)^{1/3}}\right) = (a + b) + (b - a) = 2b = 2y^{1/3}$

$= 2 \times (216)^{1/3} = 2 \times (6)^{(3\times1/3)} = 2 \times 6 = 12$

The correct answer is Option E.

13. We observe that all the options for the value of $\left(\frac{1}{\sqrt{m}}\right)$ are greater than 1. This is possible only if $\sqrt{m}$ is less than 1, i.e., m is less than 1.

Thus, we have

For $0 < m < 1$, Multiplying m throughout:

$=> 0 < m^2 < m$

Taking square root throughout:

$=> 0 < m < \sqrt{m}$

Thus, we have

$\sqrt{m} > m$

Taking reciprocal:

$\frac{1}{\sqrt{m}} < \frac{1}{m} \ldots \text{(i)}$

We know that

$m \geq 0.9$

$=> \frac{1}{m} \leq \frac{1}{0.9} = 1.11 \ldots \text{(ii)}$

Thus, from (i) and (ii), we have

$\frac{1}{\sqrt{m}} < 1.11$

Only option A (1.01) is qualified in the range.

The correct answer is Option A.

Alternate approach:

Say $m = 1$,

$=> \left(\frac{1}{\sqrt{m}}\right) = \left(\frac{1}{\sqrt{1}}\right) = 1$

We see that as the value of m increases, the value of $\left(\frac{1}{\sqrt{m}}\right)$ decreases; however, in the options, we see that there is no less than 1 value.

Since all the values in options are more than 1, we need to probe further.

If we plug in a value for m, lying between 1 and 0.9, the value of $\left(\frac{1}{\sqrt{m}}\right)$ would be greater than 1. Although all the option values are greater than 1, this does not mean that the problem can't be solved.

Since this is a question of MCQ category and only one among the five options is correct, at least Option A (1.01, least among all the options) must be correct.

14. We have $a = \frac{b^3}{90}$

=> $b^3 = 90a$

$90 = 2 \times 3^2 \times 5$

Given that b is an integer, $90a$ must be a perfect cube, we have

$a = 2^2 \times 3 \times 5^2 \times k^3$, where k is a positive integer.

In that case, we have

$90a = 2^3 \times 3^3 \times 5^3 \times k^3$, which is a perfect cube.

Thus, among the statements I, II and III, only those would be an integer in which the denominator is a factor of $a = 2^2 \times 3 \times 5^2 = 300$.

Only in Statement III, in which the denominator is 300, a is completely divisible by the denominator and hence would be an integer.

The correct answer is Option C.

15. Given that,

	X	Y
+	Y	X
X	X	Z

In the addition in the units digits, we have

$Y + X = Z +$ carry

(We must have 'carry' since in the tens position, the same addition ($X + Y$) results in a different value (X))

The value of the 'carry' must be 1 (adding two single digits can result in a maximum carry of '1')

We observe that in the addition of the tens digits, i.e. $(1 + X + Y)$, we get a 'carry' to the hundreds position (since the result is a three-digit number).

Also, we have

$(1 + X + Y)$ results in the digit X in the tens position

=> $1 + X + Y = 10 + X$

=> $1 + Y = 10$

(Since only $(X + 10)$ would result is the digit X in the tens position)

=> $Y = 9$

Since the hundreds position in the sum comes only from the 'carry' from the tens position, we have

$X = 1$

Thus, the correct addition is:

$$\begin{array}{r} 1\ 9 \\ +\ 9\ 1 \\ \hline 1\ 1\ 0 \\ \hline \end{array}$$

Thus, the value of the units digit of the integer XXZ, i.e. $Z = 0$.

The correct answer is Option E.

Alternate approach:

Since XY and YX are two digit positive integers, we must have:

$XY + YX = XXZ \leq 198$; since a two digit positive integer ≤ 99

=> X must be equal to 1

=> $XXZ = 11Z$, where Z is the units digit

We can write a two digit positive number with digits X and Y as:

$XY \equiv 10X + Y$

Similarly, we have, the two digit positive number: $YX \equiv 10Y + X$

Thus, we have

$XY + YX \equiv (10X + Y) + (10Y + X) = 11(X + Y)$

We know $XY + YX = XXZ$

=> $XXZ = 11Z \equiv 11(X + Y)$

Thus, XXZ is a multiple 11, and also less than or equal to 198:

=> $XXZ \equiv 110$

=> $Z = 0$

16. Amount saved by Suzy in the 1st month = \$20 = 20×1 = \$20.

Amount saved by Suzy in the 2nd month = \$(20 + 20) = \$40 = 20×2 = \$40.

Amount saved by Suzy in the 3rd month = \$(40 + 20) = \$60 = 20×3 = \$60.

Thus, the amount saved in the 30th month = 20×30 = \$600.

Thus, the average amount saved every month

$$= \frac{(\text{Amount saved in the } 1^{st} \text{ month}) + (\text{Amount saved in the } 30^{th} \text{ month})}{2}$$

$$= \$\left(\frac{20 + 600}{2}\right) = \$310$$

Thus, total amount saved in 30 months

= (Average amount saved per month) × (Number of months)

$= \$310 \times 30$

$= \$9,300$

The correct answer is Option D.

Alternate approach:

Sum of first n positive integers $= \frac{n \times (n+1)}{2}$

Thus, the sum of first 30 positive integers $= \frac{30 \times 31}{2} = 15 \times 31 = 465$

Thus, total amount saved in 30 months = $\$20 \times 465 = \$9,300$.

17. We have $t_n = t_{(n+1)} + 2t_{(n-1)}$ for $n \geq 1$

Since the relationship $t_n = t_{(n+1)} + 2t_{(n-1)}$ is among three terms, to get the value of t_4, we must have the values of t_2 and t_3. But we do not have the value of t_3. So, we will first calculate the value of t_3.

Plugging-in $n = 2$ in $t_n = t_{(n+1)} + 2t_{(n-1)}$, we get

$t_2 = t_3 + 2t_1$

Plugging-in the values of $t_1 = 0$, and $t_2 = 2$, we get

$2 = t_3 + 2 \times 0$

$t_3 = 2$

Again, plugging-in $n = 3$ in $t_n = t_{(n+1)} + 2t_{(n-1)}$, we get

$t_3 = t_4 + 2t_2$

Plugging-in the values of $t_2 = 2$, and $t_3 = 2$, we get

$2 = t_4 + 2 \times 2$

$t_4 = -2$

The correct answer is Option A.

18. Let us check the remainders when 2^n is divided by 3, for $n = 1, 2, 3, 4, 5\ldots$:

n	2^n	Remainder when 2^n is divided by 3
1	2	2
2	4	1
3	8	2
4	16	1
5	32	2

> Taking 2 terms: $2 + 2^2$: Remainder when $2 + 2^2$ divided by 3 = 2 + 1 = 3 => 0
> Taking 3 terms: $2 + 2^2 + 2^3$: Remainder when $2 + 2^2 + 2^3$ is divided by 3 = 2 + 1 + 2 = 5 => 2
> Taking 4 terms: $2 + 2^2 + 2^3 + 2^4$: Remainder when $2 + 2^2 + 2^3 + 2^4$ is divided by 3 = 2 + 1 + 2 + 1 = 6 => 0
> Taking 5 terms: $2 + 2^2 + 2^3 + 2^4 + 2^5$: Remainder when $2 + 2^2 + 2^3 + 2^4 + 2^5$ is divided by 3 = 2 + 1 + 2 + 1 + 2 = 8 => 2

Note that when the number of terms is even (2 and 4), the remainder is 0 and when the number of terms is odd (3 and 5), the remainder is 2.

Thus, when the number of terms ($n > 9$) is even (10, 12, ...), the remainder would be 0 and when the number of terms ($n > 9$) is odd (11, 13, ...), the remainder would be 2.

Thus, statements I and III are true.

The correct answer is Option D.

4.2 Percents

19. The cost of repairing the current machine = \$1,200

The cost of new machine = \$2,800

Since the new machine lasts for two years, the average cost per year = \$ $\left(\frac{2,800}{2}\right)$ = \$1,400.

Thus, the required percentage = $\frac{1,400 - 1,200}{1,200} \times 100 = 16.67\%$.

The correct answer is Option C.

20. Let the price of the item be \$$x$.

So tax is applicable on \$$(x - 200)$

Thus, tax paid = 10% of \$ $(x - 200)$

=> \$ $\left(\frac{10}{100} \times (x - 200)\right)$

=> \$ $\left(\frac{x - 200}{10}\right)$, which is equals to \$10.

Thus, we have

$\frac{x - 200}{10} = 10$

=> $x = 300$

The correct answer is Option C.

Alternate approach:

We see that tax paid = 10% of excess amount = \$10

=> Excess amount = \$100

Thus, the price = \$(200 + 100) = \$300

21. Tax paid on \$25 = \$0.54

Thus, a tax, which is four times as much as the above, would be \$$(0.54 \times 4)$ = \$2.16 on \$25.

Thus, the tax for item B on \$100 = 4 × Tax paid on \$25

= \$2.16 × 4 = \$8.64

Thus, this tax, expressed as a percentage = $\frac{8.64}{100} \times 100 = 8.64\%$.

The correct answer is Option C.

22. Cyclist P increased his speed from 10 mph to 25 mph.

Total increase in speed of Cyclist P = 25 – 10 = 15 mph

Thus, the percentage increase in speed of Cyclist P = $\frac{15}{10} \times 100 = 150\%$.

Cyclist Q increased his speed from 8 mph to 24 mph.

Total increase in speed of Cyclist Q = 24 – 8 = 16 mph

Thus, the percentage increase in speed of Cyclist Q = $\frac{16}{8} \times 100 = 200\%$.

Apparently it seems that the required answer is simply: 200% – 150% = 50%.

However, it is not so since we are here asked to find percent change not percent point change

Here absolute change is 200% – 150% = 50%, which we are going to compare with percent change with the speed of Cyclist P, which is 150%.

The required percent = $\frac{50}{150} \times 100 = 33.33\%$.

The correct answer is Option A.

23. Kilometers traveled by Carrier X = 101,098 =≈ 100,000 kilometers

Fuel consumed by Carrier X = 9,890 liters =≈ 10,000 liters.

Thus, mileage of Carrier X = $\frac{100,000}{10,000} = 10$ kmpl

Kilometers traveled by Carrier Y = 203,000 =≈ 200,000 kilometers

Fuel consumed by Carrier Y = 24,896 liters =≈ 25,000 liters

Thus, mileage of Carrier Y = $\frac{200,000}{25,000} = 8$ kmpl

Thus, the percent by which mileage of Carrier X is greater than that of Carrier Y

$= \left(\frac{10-8}{8}\right) \times 100$

= 25%

The correct answer is Option B.

24. Since we need to find the overall percent change, we can assume the list price equal to \$100 (the overall percent change does not depend on the list price of the bicycle).

Price after the price was decreased by 25%

$= (100 - 25)\,\%$ of \$100

$= \$100 \times \left(\frac{75}{100}\right)$

$= \$75$

Price after the decreased price was increased by 25%

$= (100 + 25)\,\%$ of \$75

$= \$75 \times \left(\frac{125}{100}\right)$

$= \$93.75$

Since the assumed list price is \$100, the final price would be 93.75% of the assumed list price. We need not take the actual list price of the bicycle into the consideration.

The correct answer is Option D.

Alternate approach:

We can find the overall percent change using the relation:

$\left(x + y + \frac{xy}{100}\right)\%$, where $x\%$ and $y\%$ represent successive percent changes.

Applying in this problem, we get:

$\left(25 - 25 - \frac{25 \times 25}{100}\right) = -6.25\%$

Thus, the overall percent change is 6.25% and the final price = 93.75% of the list price.

25. The mixing for 68 liters of base was 3.4 liters of red color.

The recommended mixing for every 10 liter of base was 0.7 liters of red color.

Thus, as per the recommendation, the amount of red color required for 68 liters of base $= \frac{0.7}{10} \times 68 = 4.76$ liters

Thus, the mixing quantity must increase by $(4.76 - 3.4) = 1.36$ liters

Thus, the percent change needed in the mixing $= \frac{1.36}{3.4} \times 100 = 40\%$

The correct answer is Option C.

26. Number of employees who worked for 30 hours in the week = 30% of 200 = 60.

Rate of pay per hour = \$7.50.

Thus, pay per employee for 30 hours = $\$(7.50 \times 30) = \$225$.

Thus, total pay for all 60 employees = $\$\ (225 \times 60) = \$13,500 \ldots$ (i)

Number of employees who worked for 44 hours in the week = 40% of 200 = 80.

Rate of pay per hour = \$7.50.

Thus, pay per employee for 44 hours= $\$(7.50 \times 44) = \$330$.

Thus, total pay for all 80 employees = $\$(330 \times 80) = \$26,400 \ldots$ (ii)

Number of employees who worked for 50 hours in the week = $200 - (60 + 80) = 60$.

Rate of pay is \$7.50 per hour for the first 44 hours and $\$\left(7.50 \times 1\frac{1}{3}\right) = \$\left(7.50 \times \frac{4}{3}\right) = \$10$ per hour for the remaining $(50 - 44) = 6$ hours.

Thus, remuneration per employee = $\$(7.50 \times 44 + 10 \times 6) = \$390$.

Thus, total remuneration for all 60 employees = $\$(390 \times 60) = \$23,400 \ldots$ (iii)

Thus, total remuneration for all the 200 employees
$= \$(13,500 + 26,400 + 23,400) = \$63,300$.

The correct answer is Option C.

27. On the first \$35 million in sales, amount received in commission = \$5 million

Thus, ratio of commission to sales = $\frac{5}{35} = \frac{1}{7}$

On the next \$121 million in sales, amount received in commission = \$11 million

Thus, ratio of commission to sales = $\frac{11}{121} = \frac{1}{11}$

Since $\frac{1}{7} > \frac{1}{11}$, there is a decrease in commission.

Thus, the required percent decrease

$$= \left(\frac{\frac{1}{7} - \frac{1}{11}}{\frac{1}{7}}\right) \times 100\%$$

$= \frac{11 - 7}{11} \times 100\%$

$= \frac{4}{11} \times 100$

$= 36.36\%$

The correct answer is Option C.

28. Total sales = \$24,000.

Thus, commission on the first \$20,000 = 8% of \$20,000 = $\$\left(\frac{8}{100} \times 20,000\right) = \$1,600$.

Total commission received by the sales representative = \$2,000 (given).

Thus, commission received on the remaining sales = $\$(2,000 - 1,600) = \$400$.

Commission received on the remaining $\$(24,000 - 20,000) = \$4,000$ at $x\%$

$= x\%$ of \$4,000

$= \$\left(\frac{x}{100} \times 4,000\right)$

$= \$40x$.

Thus, we have

$40x = 400$

$=> x = 10\%$

The correct answer is Option D.

29. Total cost of the 120,000 computer chips = \$3,600,000

Cost of $\frac{2}{5}$ of the above computer chips = $\$\left(3,600,000 \times \frac{2}{5}\right) = \$1,440,000$.

These were sold at a 25% higher than the cost price.

Thus, selling price of the above computer chips

= (100 + 25)% of \$1,440,000

$= \$\left(\frac{125}{100} \times 1,440,000\right)$

$= \$1,800,000 \ldots$ (i)

Cost of the remaining computer chips = $\$(3,600,000 - 1,440,000) = \$2,160,000$

Later, these remaining computer chips were sold at a 25% lower than the cost price.

Thus, selling price of the above computer chips

= (100 – 25)% of $2,160,000

$= \$\left(\frac{75}{100} \times 2,160,000\right)$

$= \$1,620,000 \ldots$ (ii)

Thus, total selling price = $\$(1,800,000 + 1,620,000) = \$3,420,000$.

Since total selling price (= $3,420,000) < total cost price (= $3,600,000), there is a loss

Thus, percent loss

$$= \left(\frac{\text{Cost price} - \text{Selling price}}{\text{Cost price}}\right) \times 100$$

$$= \left(\frac{3,600,000 - 3,420,000}{3,600,000}\right) \times 100$$

= 5% (Loss)

The correct answer is Option B.

Alternate approach:

Percent profit made on $\frac{2}{5}$ of the stock = 25%

Percent loss made on the remaining $(1 - \frac{2}{5} =)\frac{3}{5}$ of the stock = −25% (since price is 25% less)

Thus, overall percent profit or loss

$$= \left(\frac{2}{5} \times 25 + \frac{3}{5} \times (-25)\right)$$

= 10 – 15

= −5%

30. We know that the price of milk increased by 20%.

Let the original price of milk per liter be $\$x$.

Thus, the price of milk per liter after the price increase

$= \$\,(100 + 20)\,\%$ of $x = \$(120\%$ of $x) = \$\left(\frac{6x}{5}\right)$.

Number of liters of milk = $\left(\frac{\text{Total Price}}{\text{Price per liter}}\right)$

Thus, difference in quantity of milk obtained for \$60

$$= \frac{60}{x} - \frac{60}{\frac{6x}{5}} = 5$$

$$=> \frac{60}{x} - \frac{50}{x} = 5$$

$$=> \frac{10}{x} = 5$$

$$=> x = 2$$

Thus, the correct answer is Option A.

Alternate approach:

If the price of an item goes up/down by $x\%$, the quantity consumed should be reduced/increased by $\left(\frac{100x}{100 \pm x}\right)\%$ so that the total expenditure remains the same.

Since the price of milk increased by 20%, the quantity obtained for \$60 would reduce by $100 \times \left(\frac{20}{100 + 20}\right) = \frac{50}{3}\%$

Thus, $\frac{50}{3}\%$ of the original quantity = 5 liters

$$=> \text{Original quantity} = \frac{5}{\frac{50}{3}\%} = \frac{500}{\frac{50}{3}} = 30 \text{ liters}$$

Thus, the initial price per liter of diesel = $\left(\frac{\text{Total Initial Price}}{\text{Total Initial number of liters}}\right) = \$\left(\frac{60}{30}\right) = \$2$.

31. Since the problem asks us to find a percent value, we can assume any suitable value of the annual sum of money for ease of calculation as the initial value does not affect the final answer.

Since we have to deal with factions $\frac{1}{4}$ and $\frac{1}{6}$, we can assume the sum equal to \$24 (= LCM of 4 and 6).

Thus, the sum spent during the first quarter = $\$\left(\frac{1}{4} \times 24\right) = \$6$.

Amount of money left = $\$\,(24 - 6) = \$18$.

Thus, the sum spent during the second quarter = $\$\left(\frac{1}{6} \times 18\right) = \$3$.

Thus, the sum left at the beginning of the third quarter = $\$\,(18 - 3) = \$15$.

Total the sum spent in the last two quarters = \$ (6 + 3) = \$9.

Thus, the required percent

$= \frac{15 - 9}{9} \times 100$

$= \frac{200}{3}\%$

= 66.66%

The correct answer is Option D.

32. Amount David spent in 2013 = \$450.

Amount David spent in 2014

= 10% more than what he spent in 2013

= (100 + 10)% of what he spent in 2013

$= \$\left(\frac{110}{100} \times 450\right)$

= \$495

Total amount spent by David and Suzy in 2014 = \$600.

Thus, amount spent by Suzy in 2014 = \$ (600 – 495) = \$105.

We know that Suzy had spent \$450 in 2013.

Thus, the required percent

$= \frac{(\text{Amount spent by Suzy in 2013}) - (\text{Amount spent by Suzy in 2014})}{(\text{Amount spent by Suzy in 2013})} \times 100\%$

$= \frac{450 - 105}{450} \times 100\%$

$= \frac{345}{450} \times 100\%$

$= \approx 77\%$

The correct answer is Option C.

33. It is easier to solve this question can be solved by observing the answer options than by actual solving.

We observe that the options are very large in value compared to the price change by the end of day 2, i.e. \$1.

Let the original price of the item be $\$x$.

By the end of day 2, the price of the item decreases by \$1

Thus, by the end of day 2, the price of the item = $\$(x - 1)$

The subsequent increase would be calculated on $\$(x - 1)$ instead of $\$x$

However, since \$1 is negligible compared to x (note that x is one among very large option values), we can conclude that the decrease in price would be very slightly less than \$1.

Thus, the final price on day 4 $=\approx \$\,((x - 1) - 1) = \$(x - 2)$.

Thus, we have

$(x - 2) =\approx 398$

$=> x =\approx 400$

Thus, the only option that satisfies is D: \$400.

The actual calculation is shown below:

Assuming the initial price of a share to be $\$x$, we have

$$x - \left[x\left(1 + \frac{k}{100}\right)\left(1 - \frac{k}{100}\right)\right] = 1 \ldots \text{(i)}$$

$$x\left(1 + \frac{k}{100}\right)\left(1 - \frac{k}{100}\right)\left(1 + \frac{k}{100}\right)\left(1 - \frac{k}{100}\right) = 398 \ldots \text{(ii)}$$

We need to solve for x from the above two equations.

Note: The actual solution is time taking and very involved, and hence, not suggested.

The correct answer is Option D.

34. Cost of x items = $\$y$

Since the cost increases by 20%, the new cost of x items

$= \$\,(100 + 20)\,\%$ of y

$= \$ \left(\frac{120}{100} \times y\right)$

$= \$ \left(\frac{6y}{5}\right)$

Thus, with a budget of $\$ \left(\frac{6y}{5}\right)$, x items can be bought next year.

Thus, with a budget of $\$(3y)$, the number of items can be bought

$= \frac{3y}{\left(\frac{6y}{5}\right)} \times x$

$= \frac{5}{2}x = 2.50x$

The correct answer is Option C.

35. Since the question asks for a percent value, we can choose any suitable initial value of the total weight of the solution for the ease of calculation; the initial value will not affect the final answer.

Let the total weight of the solution initially = 100 liters.

Thus, weight of water = 30% of 100 = 30 liters.

Weight of the remaining solution = 100 – 30 = 70 liters.

Loss of water = 70% by weight.

Thus, final weight of water after 15 minutes of boiling = (100 – 70)% of 30 = 30% of 30 = 9 liters.

Since there is no weight loss in the other part of the solution, final weight of the solution = 70 + 9 = 79 units.

Thus, the required percent value = $\left(\frac{\text{Final weight of water}}{\text{Final weight of solution}}\right) \times 100$

$= \frac{9}{79} \times 100$

$= \frac{900}{79}\%$

The correct answer is Option D.

36. Since the question asks about a percent value, we can choose any suitable initial value of the total volume of the mix for ease of calculation; the initial value will not affect the final answer.

Let the total volume of the mixed juice be 100 units.

Thus, volume of banana pulp = 25% of 100 = 25 units

=> Volume of papaya pulp = 100 − 25 = 75 units

Let the price of 100 units (an equal quantity) of banana pulp be $\$100x$.

Since the mixed juice costs 20 percent more than the cost of an equal quantity of only banana pulp, the price of 100 units of the mixed juice

$= \$((100 + 20)\,\%\text{ of }100x)$

$= \$120x$

The price of 25 units of banana pulp present in the mix = $\$\left(\frac{100x}{100} \times 25\right) = \$25x$.

Thus, the price of 75 units of papaya pulp present in the mix = $\$\,(120x - 25x) = \$95x$.

Thus, the price of 100 units of papaya pulp = $\$\left(\frac{95x}{75} \times 100\right)$

$= \$126.67x$

Thus, the percent by which papaya pulp are more expensive than banana pulp

$$= \frac{(\text{Price of 100 units of papaya pulp}) - (\text{Price of 100 units of banana pulp})}{\text{Price of 100 units of banana pulp}} \times 100$$

$$= \frac{126.67x - 100x}{100x} \times 100$$

$= 26.67\%$

The correct answer is Option B.

37. Let us assume the number of days in a year to be 365.

Thus, in a 3-year period, total number of days = $3 \times 365 = 1,095$.

We know that bacteria P multiplies itself in every 18 days and bacteria Q multiplies itself in every 15 days.

Thus, we have

Number of times bacteria P multiplies itself

$= \frac{1,095}{18} = 60\ldots = 60$; bacteria P multiplies only 60 times.

Number of times bacteria Q multiplies itself

$= \frac{1,095}{15} = 73$; bacteria Q multiplies 73 times.

Since there is no information whether the 3-year period includes a leap year (366 days in the year), we must consider that scenario too.

In that case,

Number of times bacteria P multiplies itself

$= \frac{1,096}{18} = 60... = 60$; bacteria P still multiplies only 60 times.

Number of times bacteria Q multiplies itself

$= \frac{1,096}{15} = 73... = 73$; bacteria Q still multiplies only 73 times.

Even if we had not considered the leap year scenario, we would not have made any mistake. Since this is PS problem, and there cannot be two answers to the same problem.

Thus, the required percent difference

$$= \frac{\text{(Number of times bacteria Q multiplies)} - \text{(Number of times bacteria P multiplies)}}{\text{Number of times bacteria P multiplies}} \times 100\%$$

$$= \frac{73 - 60}{60} \times 100\%$$

$$= \approx 22\%$$

The correct answer is Option D.

38. Cost of the phone purchased by Jack = \$1,500.

Sales tax paid = 5% of \$1500 = $\$ \left(\frac{5}{100} \times 1,500\right) = \$75$.

Thus, total price paid by Jack = $\$ (1,500 + 75) = \$1,575$.

Cost of the phone purchased by Tom = \$1,200.

Sales tax paid = 15% of \$1,200 = $\$ \left(\frac{15}{100} \times 1,200\right) = \$180$.

Thus, total price paid by Tom = $\$ (1,200 + 180) = \$1,380$.

Thus, the amount which Tom paid less compared to Jack = $\$ (1,575 - 1,380) = \$195$.

Thus, the required % value

$$= \left(\frac{\text{Total price paid by Jack} - \text{Total price paid by Tom}}{\text{Total price paid by Jack}}\right) \times 100\%$$

$$= \left(\frac{195}{1,575}\right) \times 100\%$$

$$= \frac{260}{21} = \approx 12\%$$

The correct answer is Option D.

39. Let the number of boys and girls be b and g, respectively.

Number of boys who play basketball = 65% of b

$$= \frac{65}{100} \times b$$

$$= \frac{13b}{20}$$

Number of girls who play basketball = 78% of g

$$= \frac{78}{100} \times g$$

$$= \frac{39g}{50}$$

Thus, total number of students play basketball = $\left(\frac{13b}{20} + \frac{39g}{50}\right)$

Total number of students = $(b + g)$

Since 72% of all students play basketball, we have

$$\frac{72}{100} \times (b + g) = \frac{13b}{20} + \frac{39g}{50}$$

Upon solving, we get,

$$=> \frac{g}{b} = \frac{7}{6}$$

The correct answer is Option B.

Alternate approach:

We can solve the problem using the method of alligation, as shown below:

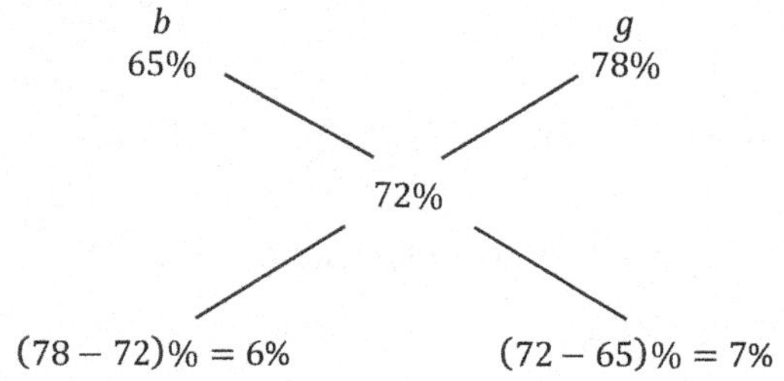

Thus, we have

$b : g = 6 : 7$

$g : b = 7 : 6$

40. Total number of spectators = S

Since 40% of the spectators belonged to natives, percent of other than native spectators = $(100 - 40) = 60\%$

Thus, the number of other than native spectators

= 60% of S

= $0.6S$

Of the above, Royal Challengers had a support of 10%.

Thus, the number of other than native supporters
= 10% of $0.6S$
= $0.06S$

Number of native supporters = 24,500

Thus, total number of supporters = $0.06S + 24,500$

The correct answer is Option D.

41. Since the problem asks for a percent value, we can assume any suitable value of the number of students for ease of calculation.

Let the number of students be 100.

Number of students of Science stream = 40% of 100 = 40.

Number of Science stream students who go to special classes
= 60% of 40
= 24

Total number of students who go to special classes = 30% of 100 = 30.

Thus, the number of students who do not study Science but go to special classes
= 30 – 24
= 6

Thus, the percent of the students who do not study Science but go to special classes
$= \frac{6}{100} \times 100$
= 6%

The correct answer is Option A.

42. Let the total number of students = 100 (since the question asks for ratio, the answer is independent of the initial number chosen, so we choose a suitable number for ease of calculation).

The number of boys = 40% of 100 = 40.

Thus, the number of girls = 100 − 40 = 60.

Let the number of students transferred = x.

Thus, number of boys transferred = 30% of $x = \frac{30}{100} \times x = \frac{3x}{10}$

Thus, number of girls transferred = $\left(x - \frac{3x}{10}\right) = \frac{7x}{10}$

Here we know that:

Transfer rate for students of a certain gender = $\frac{\text{Number of students of that gender transferred}}{\text{Total number of students of that gender}}$

Thus, the transfer rate for the boys = $\left(\frac{\frac{3x}{10}}{40}\right) = \frac{3x}{400}$.

Also, the transfer rate for the girls = $\left(\frac{\frac{7x}{10}}{60}\right) = \frac{7x}{600}$.

Thus, the required ratio = $\frac{3x}{400} : \frac{7x}{600}$

$= \frac{3}{2} : \frac{7}{3}$

$= 9 : 14$

The correct answer is Option D.

43. Since the problem asks us about a percent change, we can assume a suitable initial value of the original price for ease of calculations.

We see that we need to take $\frac{5}{7}$ and $\frac{3}{5}$ of the original price.

So, we should assume a value, which is a multiple of 35 (LCM of denominators, 7 and 5) for ease of calculations.

Thus, let the original price be \$35.

Thus, the value of the car at the start of the year = \$ $\left(\frac{5}{7} \times 35\right)$ = \$25.

The value of the car in the end of the year = \$ $\left(\frac{3}{5} \times 35\right)$ = \$21.

Thus, the required percent decrease

$= \left(\frac{\text{Change in value}}{\text{Initial value}}\right) \times 100\%$

$= \frac{25 - 21}{25} \times 100$

$= 16\%$

The correct answer is Option B.

Alternate approach:

Say the original price is $1.

Thus, the required percent decrease

$= \left(\frac{\text{Change in value}}{\text{Initial value}}\right) \times 100$

$= \frac{\frac{5}{7} - \frac{3}{5}}{\frac{5}{7}} \times 100$

$= \frac{\frac{4}{35}}{\frac{7}{5}} \times 100$

$= \frac{4}{25} \times 100$

$= 16\%$

44. Remuneration as per the first offer = 5% commission on sales plus monthly bonus of $500

Remuneration as per the second offer = 7% of commission on sales

Thus, we have

(5% commission on sales) + ($500) = (7% commission on sales)

=> (7% commission on sales) – (5% commission on sales) = $500

=> 2% of sales = $500

=> Sales = $\$ \left(500 \times \frac{100}{2}\right) = \$25,000$

The correct answer is Option B.

45. Number of retailers at the beginning of the year = 35% of 120 = $\frac{35}{100} \times 120 = 42$.

Number of retailers after the 24-month period = 25% of 240 = $\frac{25}{100} \times 240 = 60$.

Thus, increase in the number of retailers = 60 − 42 = 18.

Let the simple annual percent growth rate in the number of retailers be r%.

Thus, in two years (the 24-month period), increase in the number of retailers at r% rate

$= \left(\frac{42 \times r \times 2}{100}\right)$

(The value is calculated on 42 since 42 is the value at the start of the year)

Thus, we have

$$\frac{42 \times r \times 2}{100} = 18$$

$$=> r = \frac{18 \times 100}{2 \times 42} = 21.43\%$$

The correct answer is Option B.

46. Since we need to find the greatest overall percent increase, let us assume a suitable initial value.

Let the initial value be 100.

Let there be x% increase initially.

Thus, value after x% increase

$= (100 + x)\%$ of 100

$= \frac{100 + x}{100} \times 100$

$= (100 + x)$

Let there be y% increase next.

Thus, the final value after y% increase

$= (100 + y)\%$ of $(100 + x)$

$= \frac{(100 + y)}{100} \times (100 + x)$

$$= \frac{(100 + y)\,(100 + x)}{100}$$

$$= \frac{100^2 + 100\,(x + y) + xy}{100}$$

Thus, overall percent increase

$$= \frac{\text{(Final value)} - \text{(Initial value)}}{\text{(Initial value)}} \times 100\%$$

$$= \frac{\left(\dfrac{100^2 + 100\,(x + y) + xy}{100} - 100\right)}{100} \times 100\%$$

$$= \left(\frac{100\,(x + y) + xy}{100}\right)\%$$

$$= \left(x + y + \frac{xy}{100}\right)\%$$

In each of the above options, we observe that:

$x + y = 60$

Thus, the option with the highest value of xy will have the highest overall percent increase.

We know that if the sum of two terms is a constant, the product of the two terms becomes the maximum only when both terms are equal

$=> x = y = \frac{60}{2} = 30$

Thus, the correct answer is Option C.

Alternately, working with the options one at a time:

- A: $x = 10, y = 50 => xy = 500$
- B: $x = 25, y = 35 => xy = 875$
- C: $x = 30, y = 30 => xy =$ **900**: Correct answer.
- D: $x = 40, y = 20 => xy = 800$
- E: $x = 45, y = 15 => xy = 675$

47. Since the problem asks us about a percent value, we can assume a suitable value of the number of voters for ease of calculations.

Let the number of voters be 100.

Thus, the number of boys = 70% of 100 = 70

Number of girls = 100 – 70 = 30

Number of boys who would vote for John = 30% of 70 = 21.

Number of girls who would vote for John = 70% of 30 = 21.

Thus, total number of votes for John = 21 + 21 = 42.

Thus, the required percent = $\frac{42}{100} \times 100 = 42\%$.

The correct answer is Option B.

48. We see some ugly numbers to deal with.

Region-wise distribution of companies in the state	
Regions	**No. of companies**
Region P	2,345
Region Q	3,456
Region R	3,421
Region S	5,721
Region T	3,445
Region U	80
Region V	4,532
Total	**23,000**

In the GMAT, you would be seldom asked to do calculations that consume a lot of time. Even if you see some ugly numbers to deal with, you need not put in too much time on it. Mostly, there would be a smarter approach to deal with the question.

We are asked to find out the percentage difference between the number of companies of Region S and that of Region R. Instead of calculating the individual percentages, let's save time and directly calculate the percentage difference.

It is given that the number of companies of Region S = 5,721 and the number of companies of Region R = 3,421

Thus, the difference = 5,721 − 3,421 = 2,300.

Thus, the percentage difference of number of companies of Region S and that of region R of the total

$$= \frac{2,300}{23,000} \times 100\% = 10\%.$$

The correct answer is Option B.

4.3 Profit & Loss

49. Since $\left(\frac{4}{5}\right)^{th}$ of the stock was sold, the remaining $\left(1-\frac{4}{5}\right)=\left(\frac{1}{5}\right)^{th}$ of the stock was not sold.

Thus, $\left(\frac{1}{5}\right)^{th}$ of the total stock = 100.

Thus, $\left(\frac{4}{5}\right)^{th}$ of the total stock = $4 \times 100 = 400$.

Thus, 400 items were sold each at \$3.00.

Thus, total amount received = $\$(400 \times 3) = \$1,200$.

The correct answer is Option B.

50. We know that the trader bought 900 cartons a cost of \$20 per carton.

Thus, the total cost of the cartons = $\$\,(20 \times 900) = \$18,000$

Selling price of $\left(\frac{2}{3}\text{ of } 900\right)$ or 600 cartons = $\$\,(1.25 \times 20) = \$25$ per carton

Thus, selling price of 600 cartons = $\$\,(25 \times 600) = \$15,000 \ldots$ (i)

Selling price of the remaining $(900 - 600)$ or 300 cartons = $\$\,(100 - 20)\%$ of 20 = $\$\left(\frac{80}{100} \times 20\right) = \$16$ per carton

Thus, selling price of 300 cartons = $\$\,(16 \times 300) = \$4,800 \ldots$ (ii)

Thus, from (i) and (ii):

Total selling price = $\$(15,000 + 4,800) = \$19,800$.

Thus, gross profit = Total selling price – Total cost price

$= \$(19,800 - 18,000) = \$1,800$.

The correct answer is Option A.

51. Selling price of a brand A bicycle = \$150.

Thus, 60% of the selling price of a brand B bicycle = \$150

Thus, selling price of a brand B bicycle = $\$\left(150 \times \frac{100}{60}\right) = \$250$

Total number of bicycles sold by the dealer = 100

Thus, the number of brand B bicycles sold = $\frac{3}{5} \times 100 = 60$

Thus, the number of brand A bicycles sold = $100 - 60 = 40$

Thus, total sale from the sale of all bicycles

= Sale from brand A bicycles + Sale from brand B bicycles

$= \$ (150 \times 40) + \$ (250 \times 60)$

$= \$ (6,000 + 15,000)$

$= \$21,000$

The correct answer is Option D.

52. Purchase price of the consignment = \$800.

Percent profit made on the Purchase price = 30%.

Thus, selling price of the consignment
$= (100 + 30)\%$ of \$800

$= \$ \left(\frac{130}{100} \times 800\right)$

$= \$1,040$

This selling price is 20% less than the marked price.

Thus, $(100 - 20)\% = 80\%$ of the marked price is equal to the selling price \$1,040.

Thus, the marked price

$= \$ \left(1,040 \times \frac{100}{80}\right)$

$= \$1,300$

The correct answer is Option C.

53. Cost of production of each unit = \$2.50

Selling price of each unit = \$4.50

Thus, margin on each unit = $\$ (4.50 - 2.50) = \$2.00$

Investment made on the equipment = \$10,000.

Thus, number of units required to be sold to recover the investment in machines

$= \frac{\text{Investment}}{\text{Margin on each unit}} = \frac{10,000}{2.00}$

$= 5,000$

The correct answer is Option D.

Alternate approach:

Let the number of required units o be sold to recover the investment in machines = n.

Thus, total cost = \$ $(10,000 + 2.5n)$.

Total selling price = \$$4.5n$

Thus, we have

$10,000 + 2.5n = 4.5n$

$=> n = 5,000$

54. Original selling price of the house

= $(100 + 20)\,\%$ of the cost of the house

= 120% of the cost of the house

New selling price of the house

= $(100 + 30)\,\%$ of the cost of the house

= 130% of the cost of the house

Thus, difference between the above two selling prices

= (130% – 120%) of the cost of the house

= 10% of the cost of the house

Since it is given that the difference between the two selling prices is \$10,000, we have

10% of the cost of the house = \$10,000

=> Cost of the house = $\$ \left(10,000 \times \frac{100}{10}\right) = \$100,000$

Thus, the original selling price of the house

= 120% of the cost of the house

= $\$ \left(100,000 \times \frac{120}{100}\right)$

= $120,000

The correct answer is Option C.

55. Total number of television sets assembled = 600.

Contractor payment for each of the first 100 television sets = $20

Thus, the total contractor payment for the first 100 television sets = $\$ (20 \times 100) = \$2,000$.

Contractor payment for each of the remaining 500 (= 600 – 100) television sets = $15

Thus, the total contractor payment for the remaining 500 television sets = $\$ (15 \times 500) = \$7,500$.

Thus, the total contractor payment for 600 television sets = $\$ (2,000 + 7,500) = \$9,500$.

Total invoice value for the 600 television at $25.00 each = $\$ (25 \times 600) = \$15,000$.

Thus, gross profit = Total invoice value – Total contractor payment

= $15,000 – $9,500

= $5,500

The correct answer is Option C.

56. Initial selling price of the item = $99.

Since the initial percent profit was 10% of the cost, we have

$(100 + 10)\,\%$ of the cost of the item = $99

=> Cost of the item = $\$ \left(\frac{99}{110} \times 100\right) = \$90$

Increased selling price of the item = $117.

Thus, the required percent profit

$= \left(\frac{\text{Selling price} - \text{Cost price}}{\text{Cost price}}\right) \times 100$

$= \left(\frac{117 - 90}{90}\right) \times 100$

$= \frac{27}{90} \times 100$

$= 30\%$

The correct answer is Option E.

57. The merchant has 2,400 fans priced at (cost price) \$30 each.

Thus, total cost price = $\$\,(2,400 \times 30) = \$72,000$.

Number of fans sold at \$40 each = 60% of 2,400 = $\frac{60}{100} \times 2400 = 1,440$.

Thus, sales revenue of these 1,440 fans = $\$\,(1440 \times 40) = \$57,600$.

Number of fans sold at \$35 each = 2,400 – 1,440 = 960.

Thus, sales revenue of these 960 fans = $\$\,(960 \times 35) = \$33,600$.

Thus, sales revenue of 2,400 fans = 57,600 + 33,600 = \$91,200

Thus, profit made by selling the 2,400 fans = $91,200 - 72,000 = \$19,200$.

Thus, average profit made per fan = $\$\left(\frac{19,200}{2,400}\right) = \$8$.

The correct answer is Option B.

Alternate approach:

The merchant sold 60% of the fans each at a profit of $\$\,(40 - 30) = \$10$.

He sold the remaining 40% of the fans each at a profit of $\$\,(40 - 35) = \$5$.

Thus, his average profit per radio = $\$\left(10 \times \frac{60}{100} + 5 \times \frac{40}{100}\right)$

$= \$\,(6 + 2)$

$= \$8$

58. Let us understand the problem using the diagram shown below:

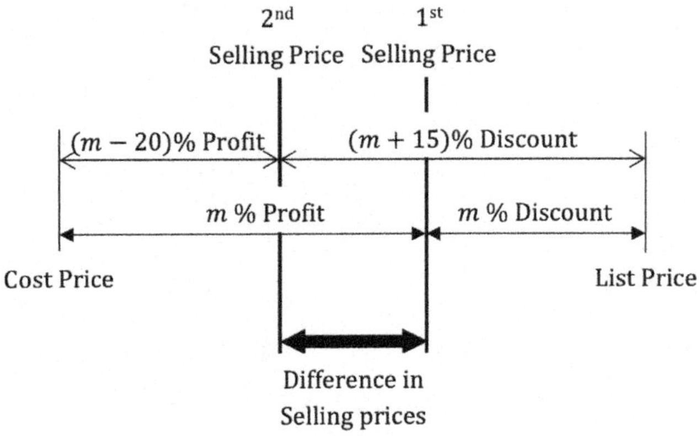

Let us calculate the difference in the two selling prices:

Since in the first case, the profit was $m\%$ and in the second, the profit was $(m-20)\%$, the difference in selling price

$= (m-(m-20))\%$ of Cost Price; (Since percent discount is calculated on the cost price)

$= 20\%$ of Cost price . . . (i)

Again, in the first case, the discount was $m\%$ and in the second, the discount was $(m+15)\%$, the difference in selling price

$= ((m+15)-m)\%$ of List price; (Since percent discount is calculated on the list price)

$= 15\%$ of List price . . . (ii)

Thus, from (i) and (ii), we have

20% of Cost price = 15% of List price

$$=> \frac{\text{List price}}{\text{Cost price}} = \frac{20}{15} = \frac{4}{3}$$

Thus, if no discount were offered (the list price becomes the selling price), percent profit

$$= \left(\frac{\text{List price} - \text{Cost price}}{\text{Cost price}}\right) \times 100$$

$$= \left(\frac{\text{List price}}{\text{Cost price}} - 1\right) \times 100$$

$$= \left(\frac{4}{3} - 1\right) \times 100$$

$= 33.3\%$

The correct answer is Option D.

59. Total number of units sold = 800 + 900 = 1,700

Cost of producing each unit = \$6

Thus, total cost of producing 1700 units = \$(6 × 1700) = \$10,200

Selling price of 800 units of the product = \$(800 × 8) = \$6,400

Selling price of 900 units of the product = \$(900 × 5) = \$4,500

Thus, the total selling price of 1,700 units = \$(6,400 + 4,500) = \$10,900

Thus, profit = $\$\ (10,900 - 10,200) = \$700$

The correct answer is Option E.

60. Let the book and stationary sales in 2014 be b and s, respectively.

Thus, total sales revenue in 2014 = $(b + s)$

Thus, the sales revenue from book sales in 2015 = (100 – 10)% of b = 90% of $b = 0.9b$.

And, the sales revenue from stationary sales in 2015 = (100 + 6)% of s = 106% of $s = 1.06s$.

Thus, total sales revenue in 2015 = $(0.9b + 1.06s)$

Thus, we have

$(0.9b + 1.06s) = (100 + 2)\,\%$ of $(b + s)$

=> $0.9b + 1.06s = 1.02b + 1.02s$

=> $0.04s = 0.12b$

=> $\frac{b}{s} = \frac{0.04}{0.12} = \frac{1}{3}$

=> $b : s = 1 : 3$

The correct answer is Option A.

61. Let the profit of the trader in 2001 be \$100 (the assumption of \$100, or any other number does not affect the answer since we have to find the percent change).

Thus, the profit of the trader in 2002 = $\$((100 + 20)\,\%$ of $100) = \$120$.

Thus, the profit of the trader in 2003 = $\$((100 + 25)\,\%$ of $120) = \$150$.

Thus, the percent change in profit from 2001 to 2003 = $\frac{150 - 100}{100} \times 100 = 50\%$.

The correct answer is Option E.

Alternate approach:

If the value of a commodity changes by $x\%$ and then by $y\%$ successively, the overall percent change is given by:

$$\left(x + y + \frac{xy}{100}\right)\%$$

Here, we have

x = Percentage change from 2001 to 2002 = 20 %

y = Percentage change from 2002 to 2003 = 25 %

Thus, the overall percent change

$$= 20 + 25 + \frac{20 \times 25}{100}$$

$$= 50\%$$

4.4 Averages

62. Number of teachers = $\frac{\text{Total annual salaries}}{\text{Average annual salaries}} = \frac{3,780,000}{42,000} = 90$

=> Number of students = $\frac{25}{2} \times 90 = 1,125$

The correct answer is Option C.

63. Let the number of male employees and female employees be m and f, respectively.

Total salary of male employees = $65,000m$.

Total salary of female employees = $80,000f$.

Thus, total salaries of all the employees = $65,000m + 80,000f$.

Thus, average salaries of all the employees

$= \left(\frac{\text{Total salaries of the employees}}{\text{Total number of employees}}\right) = \left(\frac{65,000m + 80,000f}{m+f}\right)$.

Thus, we have

$$\left(\frac{65,000m + 80,000f}{m+f}\right) = 70,000$$

$=> 65m + 80f = 70m + 70f$

$=> 10f = 5m$

$=> \frac{m}{f} = \frac{2}{1}$

We cannot get the absolute number of male employees and female employees. There are infinite number of possibilities.

Among the options, the value that bears the ratio 2 : 1 would be the correct answer. We see that only Option D (14; 7) bears that ratio 2: 1.

The correct answer is Option D.

64. It is given that the total number of students = 40.

Let the number of students in section A = n.

Thus, the number of students in section B = $(40 - n)$.

Total score of all the students in the two sections combined = $85n + 80(40 - n)$.

Thus, average score considering all students

$$= \frac{\text{Total score}}{\text{Total number of students}} = \frac{85n + 80\,(40 - n)}{40}$$

Thus, we have

$$\frac{85n + 80\,(40 - n)}{40} = 82$$

$=> 17n + 640 - 16n = 656$

$=> n = 16$

The correct answer is Option C.

65. The juice manufacturer has $(1,200 + 400) = 1,600$ liters of mango pulp.

Amount of water in the first stock = 25% of 1200 = 300 liters

Amount of water in the second stock = 20% of 400 = 80 liters

Thus, the total amount of water = 300 + 80 = 380 liters

Thus, the percent of water in the total stock

$$= \frac{380}{1,600} \times 100$$

= 23.75%.

The correct answer is Option B.

Alternate approach:

The required percent is the weighted average of the percentages of the above two stocks

$$= \left(\frac{\left(1,200 \times \frac{25}{100}\right) + \left(400 \times \frac{20}{100}\right)}{1,200 + 400}\right) \times 100$$

$$= \left(\frac{300 + 80}{1,600}\right) \times 100$$

$$= \left(\frac{380}{1,600}\right) \times 100$$

= 23.75%.

66. Let the number of students in the sections P, Q, R, and S be $p, q, r,$ and s, respectively.

Thus, the average weight of all students in the class

$$= \left(\frac{\text{Total weight of all the students combined}}{\text{Total number of students}}\right)$$

$$= \left(\frac{45 \times p + 50 \times q + 55 \times r + 65 \times s}{p + q + r + s}\right) \text{lb.}$$

Thus, we have

$$\left(\frac{45 \times p + 50 \times q + 55 \times r + 65 \times s}{p + q + r + s}\right) = 55$$

$=> 45p + 50q + 55r + 65s = 55p + 55q + 55r + 55s$

$=> 10p + 5q = 10s$

$=> 2p + q = 2s$

Since we need to maximize r, we need to find the minimum possible values of $p, q,$ and s so that the above equation holds true.

Since the RHS of the above equation is $2s$, it is even.

Also, in the LHS of the above equation, $2p$ is even.

Thus, q must be even.

The smallest even number that we can consider for q is 2 since we have at least one student in each section. Thus, we have $q = 2$.

Thus, the equation gets modified to:

$2p + 2 = 2s$

$=> p + 1 = s$

Thus, we use the minimum possible values: $p = 1, \ s = 2$.

Thus, we have $p = 1, \ q = 2$ and $s = 2$.

Since there are a total of 40 students in all sections combined, the maximum value of students in section R $= r = 40 - (p + q + s)$

$= 40 - 5 = 35$

The correct answer is Option D.

67. Since the set starts with an odd number (1) and has an odd numbers of integers, the set would end with an odd number, too. Let's see the set.

Set N: {1, 2, 3, 4, 5, ..., (2n + 1)}, where n is a positive integer.

=> Number of odd terms is one more than the number of even terms.

Thus, the number of even terms = n

The number of odd terms = $(n + 1)$

We know that the sum (M) of p terms of an arithmetic progression having first term as a and common difference as d is given by:

$$M = \frac{p}{2}(2a + (p - 1) \times d)$$

Say, X is the average of the odd integers in set N and Y is the average of the even integers in set N

Thus, we have

$$X = \frac{(1 + 3 + 5 + \cdots + (2n + 1))}{n + 1} = \frac{\left(\frac{n + 1}{2}\right) \times (2 \times 1 + ((n + 1) - 1) \times 2)}{n + 1} = n + 1$$

$$Y = \frac{(2 + 4 + \cdots + 2n)}{n} = \frac{2 \times (1 + 2 + \cdots + n)}{n} = \frac{2 \times \left(\frac{n(n + 1)}{2}\right)}{n} = n + 1$$

Thus, we have

$$X - Y = (n + 1) - (n + 1) = 0$$

The correct answer is Option B.

Alternate approach:

Since there is no restriction on the number of terms, let us try to find $(X - Y)$ using a few values of the number of terms.

Say there are only three terms in the set.

S = {1, 2, 3}

$$X = \frac{1 + 3}{2} = 2$$

$Y = 2$

=> $X - Y = 0$

Again, say there are only five terms in the set:

S = {1, 2, 3, 4, 5}

$$X = \frac{1 + 3 + 5}{3} = 3$$

$$Y = \frac{2 + 4}{2} = 3$$

$=> X - Y = 0$

Since $(X - Y) = 0$ for the number of terms = 3 and the number of terms = 5, we can conclude that the value of $(X - Y)$ would be the same for any odd number of terms in the set, as we have a definite answer among the given answer options.

68. Let the seven numbers be a, b, c, d, e, f, g, h and i.

Thus, we have

$$\frac{a + b + c + d + e + f + g + h + i}{9} = 25$$

$=> a + b + c + d + e + f + g + h + i = 225 \ldots$ (i)

Since the average of the first five numbers is 20, we have

$$\frac{a + b + c + d + e}{5} = 20$$

$=> a + b + c + d + e = 100 \ldots$ (ii)

Since the average of the last five numbers is 32, we have

$$\frac{e + f + g + h + i}{5} = 32$$

$=> e + f + g + h + i = 160 \ldots$ (iii)

Adding (ii) and (iii), we have

$a + b + c + d + 2e + f + g + h + i = 260 \ldots$ (iv)

Subtracting (i) from (iv), we have

$e = 260 - 225 = 35$

Thus, the value of the fifth number is 35.

The correct answer is Option C.

69. Let the radius of each green ball = x inches.

Each green ball is 4 inches less than the average radius of the balls in Box X.

Thus, the average radius of balls in Box X = $(x + 4)$ inches.

Also, each green ball is 2 inches greater than the average radius of the balls in Box Y.

Thus, the average radius of balls in Box Y = $(x - 2)$ inches.

Thus, the required difference = $((x + 4) - (x - 2)) = 6$ inches.

Note: In this problem, there is a lot of data which has been given to make the question appear complicated. The radius of yellow balls is of no consequence. One should carefully read the problem statement and use only the information required to answer the question.

The correct answer is Option B.

70. We know that

Average salary

$$= \frac{(\text{\# of emps of group I} \times \text{Av. salary}) + (\text{\# of emps of group II} \times \text{Av. salary}) + (\text{\# of emps of group III} \times \text{Av. salary})}{\text{Total number of employees}}$$

$$\Rightarrow \text{Average salary} = \$ \left(\frac{10 \times 35,000 + 35 \times 30,000 + 15 \times 60,000}{60}\right)$$

$= \$38,333$

The correct answer is Option B.

Note: While calculating the average, we may work with the salaries as \$35, \$30 and \$60, respectively; and after calculating the average, multiply the result with 1,000.

71. Let us use the method of alligation to solve this problem:

Total number of pencils and erasers bought = 20.

Price of each pencil = 20 cents.

Price of each eraser = 30 cents.

Average price of all 20 pieces = 28 cents.

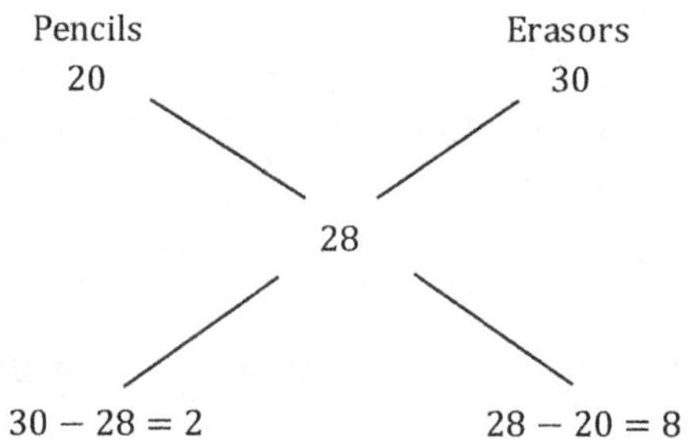

Thus, ratio of the number of pencils to the number of erasers = 2 : 8 = 1 : 4.

Since total number of pieces purchased are 20, we have

Number of pencils = $\left(\frac{1}{1+4}\right) \times 20 = 4$

Number of erasers = $\left(\frac{4}{1+4}\right) \times 20 = 16$

We now need to find the number of erasers required to be returned so that the average price falls to 26 cents.

So the situation is this,

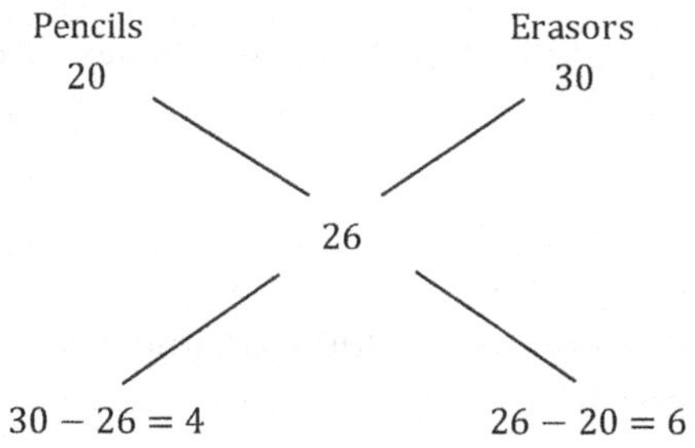

Thus, ratio of the number of pencils to the number of erasers now = 4 : 6 = 2 : 3.

Since the number of pencils remains the same from the initial condition (only erasers are returned) and we had obtained 2 pencils on ratio scale, so here too, the number of pencils should be 2 on ratio scale.

Thus, the number of erasers must be 6 (since the ratio of pencils to erasers = 2 : 3).

Since initially, there were 16 erasers, number of erasers returned = 16 − 6 = 10.

The correct answer is Option E.

Alternate approach 1:

Average amount spent for 20 pieces (pencils and erasers) = 28 cents.

Total amount spent for 20 pieces = $28 \times 20 = 560$ cents.

Let us assume that the boy returned x erasers so that the average amount spent is 26 cents.

Thus, total amount spent for $(20 - x)$ pieces = $26 \times (20 - x)$ cents.

Thus, price of x erasers returned = $(560 - 26 \times (20 - x))$ cents.

Since the price of one eraser is 30 cents, we have

$$30 = \frac{560 - 26 \times (20 - x)}{x}$$

$=> 30x = 560 - 520 + 26x$

$=> x = 10$

Alternate approach 2:

The average price of 20 pieces was 28 cents and after returning some erasers, say x, the average of $(20 - x)$ pieces became 26 cents.

Alternately, we can say that:

The average price of some pieces, say f, was 26 cents and after adding some erasers, say r (at 30 cents each), the average of $(f + r = 20)$ pieces became 28 cents.

Thus, by alligation method:

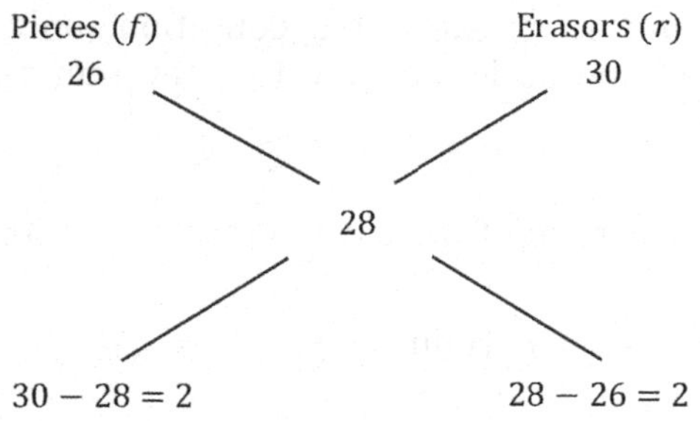

Thus, we have

$f : r = 2 : 2 = 1 : 1$

However, we know that:

$f + r = 20$

$=> f = r = \left(\frac{1}{1+1} \times 20\right) = 10$

Thus, the number of erasers returned is 10.

Alternate approach 3:

We infer the same conclusion as in the previous approach:

The average price of some pieces, say f, was 26 cents and after adding some erasers say r (at 30 cents each), the average of $(f + r = 20)$ pieces became 28 cents.

Thus, the price of each of the f pieces increased by (28 – 26) = 2 cents, resulting in a total increase of $2 \times f = 2f$ cents.

Also, the price of each of the r erasers reduced by (30 – 28) = 2 cents resulting in a total decrease of $2 \times r = 2r$ cents.

The increase of $2f$ cents came at the expense of the reduction of $2r$ cents, implying:

$2r = 2f$

$=> r = f$

But, we have

$f + r = 20$

$=> f = r = \left(\frac{1}{1+1} \times 20\right) = 10$

Thus, the number of erasors returned is 10.

72. We know that the student's average score on four tests is 78.

Thus, his total score on the four tests = $4 \times 78 = 312$.

Let the score on the 5th test be n.

Thus, his total score on the five tests = $(312 + n)$.

Thus, his average on the 5 tests = $\left(\frac{312 + n}{5}\right)$.

We know that

The final average increases from the average on 4 tests by an integer

This is possible only when the average of the 5 tests is also an integer.

Again, the student's average (an integer value) on the 5 tests = $\frac{312 + n}{5} = 62 + \frac{(2 + n)}{5}$

Thus, $(n + 2)$ must be divisible by 5.

Working with the options, we see that only Option D, i.e. $n = 93$ satisfies since $(n + 2)$ is divisible by 5.

The correct answer is Option D.

73. We know that the ratio of the numbers of candidates in groups P, Q and R was $3 : 5 : 4$, respectively.

Let the number of candidates in groups P, Q and R be $3x, 5x$ and $4x$, respectively, where x is a constant of proportionality.

We also know that the average scores for the groups P, Q and R were 64, 84, and 72, respectively.

Thus, the average score for the three groups combined

$$= \left(\frac{64 \times 3x + 84 \times 5x + 72 \times 4x}{3x + 5x + 4x}\right)$$

$$= \left(\frac{192x + 420x + 288x}{12x}\right)$$

$$= \left(\frac{900x}{12x}\right)$$

$$= 75$$

The correct answer is Option B.

Alternate approach:

Let us solve the problem using the concept of assumed mean (deviation method).

We know that the ratio of the numbers of candidates in the groups P, Q and R was $3 : 5 : 4$, respectively.

Let the number of candidates in groups P, Q and R be $3x, 5x$ and $4x$, respectively, where x is a constant of proportionality.

We know that the average scores for the groups P, Q and R were 64, 84, and 72, respectively.

Let the assumed mean score for the three groups combined = 64.

Thus, the effective scores are: $(64 - 64) = 0$, $(84 - 64) = 20$ and $(72 - 64) = 8$

Thus, we have

Average = $\frac{0 \times 3x + 20 \times 5x + 8 \times 4x}{3x + 5x + 4x}$

$= \frac{132x}{12x}$

$= 11$

Thus, the actual average = 11 + 64 = 75.

Note: This approach leads to lesser calculations than that in the previous approach.

74. We know that there are 50 male and 20 female members.

Thus, the total number of members = 50 + 20 = 70.

As the average of all 70 members is 23 years, sum of their ages = $70 \times 23 = 1,610$

As the average of 50 male members is 20 years, sum of their ages = $50 \times 20 = 1,000$

Thus, the sum of the ages of 20 female members = $1,610 - 1,000 = 610$

Thus, the average age of these 20 female members = $\frac{610}{20} = 30.50$

The correct answer is Option A.

Alternate approach 1:

Here we know that the average age of 70 members = 23 years.

The average age of 50 male members = 20

Let us consider the average age of the remaining 20 female members = x years.

By the method of alligation:

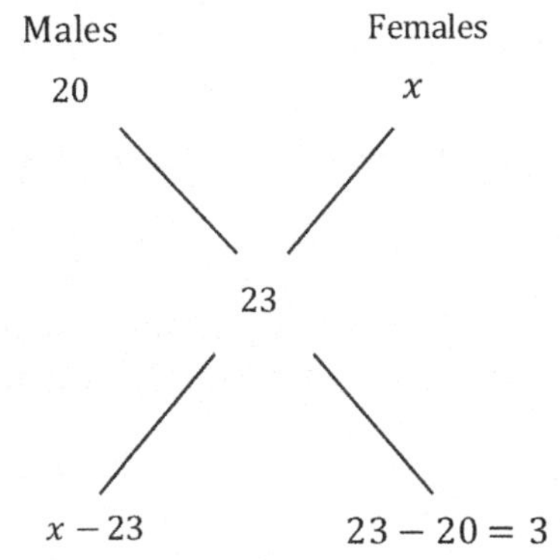

Thus, we have

$$\frac{x-23}{3} = \frac{50}{20} = \frac{5}{2}$$

$=> 2(x-23) = 5 \times 3$

$=> 2x - 46 = 15$

$=> 2x = 61$

$x = 30.50$ years.

Alternate approach 2:

According to the data, we have

Average age of 50 male members is 20 years and the average age of all 70 members is 23 years.

Let us reduce the above averages by 20.

Thus, the modified data is:

Average age of 50 male members is 0 years and the average age of all 70 members is 3 years.

Let, in the above situation, the average age of the 20 female members be a years.

Thus, we have

$$\frac{50 \times 0 + 20a}{70} = 3$$

$=> 210 = 20a$

$=> a = 10.5$

To get the actual average age of the females, we must add 20 (which we had subtracted initially).

Thus, the actual average age of the females = 10.5 + 20 = 30.5

Although practically, above approach is not preferred one, for the sake of understanding, you must learn it.

75. This question is crafted to acknowledge the importance of Alternate approach 2 discussed in the previous question. In this question, while other approaches will certainly consume more time, Alternate approach 2 would excel.

According to the data, we have

Average age of 50 male members is 20.89 years and the average age of all 50 + 20 = 70 members is 23.89 years.

Let us reduce the above averages by 20.89.

Thus, the modified data is:

Average age of 50 male members is 20.89 – 20.89 = 0 years and the average age of all 70 members is 23.89 – 20.89 = 3 years.

Let, in the above situation, the average age of the 20 female members be a years.

Thus, we have

$$\frac{50 \times 0 + 20a}{70} = 3$$

$=> 20a = 210$

$=> a = 10.50$

To get the actual average age of the females, we must add 23.89 (which we had subtracted initially).

Thus, the actual average age of the females = 10.50 + 20.89 = 31.39 years

The correct answer is Option C.

4.5 Ratio & Proportion

76. Fixed cost = \$25,000.

The total cost for 50,000 bearings = \$100,000.

Thus, the variable component of the total cost for 50,000 bearings = $\$\,(100,000 - 25,000) = \$75,000$.

Thus, variable cost per bearing = $\$\left(\frac{75,000}{50,000}\right) = \$1.50$.

Thus, the variable component of the total cost for 100,000 bearings = $\$\,(100,000 \times 1.50) = \$150,000$.

Thus, the total cost for 100,000 bearings = $\$\,(25,000 + 150,000) = \$175,000$.

The correct answer is Option C.

77. We know that the beaker was filled with 40 liters of water and liquid chemical with the components in the ratio 3 : 5, respectively.

Initial quantity of water = $\left(\frac{3}{3+5}\right) \times 40 = 15$ liters.

Initial quantity of liquid chemical = $\left(\frac{5}{3+5}\right) \times 40 = 25$ liters.

Amount of water evaporated per day = 2% of 15= 0.3 liters.

Thus, total amount of water evaporated in 10 days = $0.3 \times 10 = 3$ liters.

Amount of liquid chemical evaporated per day = 5% of 25 = 1.25 liters.

Thus, total amount of liquid chemical evaporated in 10 days = $1.25 \times 10 = 12.5$ liters.

Thus, total amount of mixture evaporated in 10 days = $3 + 12.5 = 15.5$ liters.

Thus, the percent of the original amount of mixture evaporated

$= \frac{15.5}{40} \times 100$

= 38.75%.

The correct answer is Option C.

78. Fraction of total dolls that are Barbie dolls = $\frac{3}{5}$ . . . (i)

Thus, fraction of total dolls that are non-Barbie dolls = $1 - \frac{3}{5} = \frac{2}{5}$ . . . (ii)

Fraction of Barbies purchased before the age of 10 = $\frac{4}{7}$

Thus, fraction of Barbies purchased at the age of 10 or later = $1 - \frac{4}{7} = \frac{3}{7}$ . . . (iii)

Thus, from (i) and (iii), we have

Fraction of total dolls that are Barbies and purchased at the age of 10 or later

$= \frac{3}{5} \times \frac{3}{7} = \frac{9}{35}$

Thus, we have

$\frac{9}{35}$ of the total dolls = 90

=> Total dolls = $90 \times \frac{35}{9} = 10 \times 35 = 350$

Thus, from (ii):

Number of non-Barbie dolls = $350 \times \frac{2}{5} = 140$

The correct answer is Option C.

79. We know that, for a ratio $0 < \frac{x}{y} < 1$:

- $0 < \frac{x}{y} < \left(\frac{x+k}{y+k}\right) < 1$ ... (i), if k is a positive number
- $0 < \left(\frac{x-k}{y-k}\right) < \frac{x}{y} < 1$ ... (ii)

Also, for a ratio $\frac{x}{y} > 1$:

- $1 < \left(\frac{x+k}{y+k}\right) < \frac{x}{y}$ ... (iii), if k is a positive number
- $1 < \frac{x}{y} < \left(\frac{x-k}{y-k}\right)$ ... (iv)

In the above problem, we have

The given ratio of ages of John and Suzy = $\frac{5}{6}$ $(< 1) = \approx 0.83$

Thus, after 10 years, ages of both would increase by 10.

Hence, the final ratio must be greater than $\frac{5}{6}(= 0.83)$ (from relation (i) above).

Working with the options, we have

Option A: $\frac{2}{3} = 0.67 \not> 0.83$ – Does not satisfy

Option B: $\frac{13}{20} = 0.65 \not> 0.83$ – Does not satisfy

Option C: $\frac{11}{15} = 0.73 \not> 0.83$ – Does not satisfy

Option D: $\frac{4}{5} = 0.8 \not> 0.83$ – Does not satisfy

Option E: $\frac{9}{10} = 0.90 > 0.83$ – Satisfies

The correct answer is Option E.

80. Since the problem asks us to find a fraction value, we can assume any suitable value of the total number of phones and the time taken to produce a feature phone since the initial value does not affect the final answer.

Let the total number of phones be 5.

Thus, the number of feature phones = $\frac{2}{5} \times 5 = 2$.

Number of smartphones = $5 - 2 = 3$.

Let the time taken to produce a feature phone = 5 hours.

Thus, the time taken to produce a smartphone = $\frac{8}{5} \times 5 = 8$ hours.

Thus, total time taken to produce smartphones = $3 \times 8 = 24$ hours.

Total time taken to produce feature phones = $2 \times 5 = 10$ hours.

Thus, total time taken to produce all the phones = $24 + 10 = 34$ hours.

Thus, the required fraction = $\frac{24}{34} = \frac{12}{17}$

The correct answer is Option C.

81. We have

$$\frac{\text{Number of shirts}}{\text{Number of trousers}} = \frac{4}{5} \ldots \text{(i)}$$

$$\frac{\text{Number of jackets}}{\text{Number of shirts}} = \frac{3}{8} \ldots \text{(ii)}$$

$$\frac{\text{Number of sweaters}}{\text{Number of trousers}} = \frac{6}{5}$$

Taking reciprocal on both sides:

$$=> \frac{\text{Number of trousers}}{\text{Number of sweaters}} = \frac{5}{6} \ldots \text{(iii)}$$

Thus, from the above three equations, we have

$$\frac{\text{Number of jackets}}{\text{Number of sweaters}}$$

$$= \left(\frac{\text{Number of jackets}}{\text{Number of shirts}}\right) \times \left(\frac{\text{Number of shirts}}{\text{Number of trousers}}\right) \times \left(\frac{\text{Number of trousers}}{\text{Number of sweaters}}\right)$$

$$= \frac{3}{8} \times \frac{4}{5} \times \frac{5}{6}$$

$$= \frac{1}{4}$$

The correct answer is Option C.

82. Since the question asks for a fraction value, we can choose any suitable initial value of the total number for members for ease of calculation as the initial value will not affect the final answer.

Let the number of members be 100.

Thus, the number of male members $= \frac{3}{5} \times 100 = 60$.

Number of female members $= 100 - 60 = 40$.

Fraction of male members who attended the prayer $= \frac{3}{5}$

Thus, the fraction of male members who did not attend the prayer $= \left(1 - \frac{3}{5}\right) = \frac{2}{5}$

Thus, the number of male members who did not attend the prayer $= \frac{2}{5} \times 60 = 24$.

Fraction of female members who attended the prayer $= \frac{7}{10}$

Thus, the fraction of female members who did not attend the prayer $= \left(1 - \frac{7}{10}\right) = \frac{3}{10}$

Thus, the number of female members who did not attend the prayer $= \frac{3}{10} \times 40 = 12$.

Thus, total number of members who did not attend the prayer = 24 + 12 = 36.

Thus, the required fraction = $\frac{\text{Number of male members who did not attend the prayer}}{\text{Total number of members who did not attend the prayer}}$

$= \frac{24}{36} = \frac{2}{3}$

The correct answer is Option C.

83. Total amount to be paid = \$100.

Amount paid by Suzy = \$20.

Thus, the amount paid by John and David = \$ (100 – 20) = \$80.

We know that John paid $\left(\frac{5}{3}\right)^{\text{th}}$ of what David paid.

Thus, ratio of the amounts paid by John and David = 5 : 3.

Thus, we need to divide \$80 in the ratio 5 : 3.

Thus, amount paid by David

$= \$\left(\frac{3}{5+3}\right) \times 80 = 30$

Thus, the fraction of the total amount paid by David

$= \frac{30}{100} = \frac{3}{10}$

The correct answer is Option D.

84. Total number of shirts and trousers = X.

We know that the ratio of the number of trousers to the number of shirts = 1 : 5.

Thus, the number of shirts

$= \left(\frac{5}{5+1}\right) \times X$

$= \frac{5X}{6}$

Thus, the number of cotton shirts = $\left(\frac{1}{5}\right)^{\text{th}}$ of the number of shirts

$= \frac{1}{5} \times \frac{5X}{6}$

$= \frac{X}{6}$

The correct answer is Option E.

85. Total weight of the rod = 20 pounds.

We know that the weight (w) of each piece is directly proportional to the square of its length (l).

Weight of the first piece (w_1), which is 36 feet long (l_1) = 16 pounds.

Thus, weight of the other piece (w_2) = 20 − 16 = 4 pounds.

Thus, we have

$w \propto l^2$

$=> \frac{w_1}{w_2} = \frac{(l_1)^2}{(l_2)^2}$

$=> \frac{16}{4} = \frac{36^2}{(l_2)^2}$

$=> (l_2)^2 = 36^2 \times \left(\frac{4}{16}\right)$

$=> (l_2)^2 = \frac{36^2}{4}$

$=> (l_2)^2 = \left(\frac{36}{2}\right)^2$

$=> (l_2)^2 = 18^2$

$=> (l_2)^2 = 18$

Thus, the length of the second piece = 18 feet.

The correct answer is Option C.

86. We know that the ratio of coins of John to that of Suzy = 3 : 4

Let John's and Suzy's coins be $3x$ and $4x$, respectively, where x is a constant of proportionality.

Thus, the total number of coins = $3x + 4x = 7x$.

Since John's share exceeds $\frac{2}{7}$ of the total number of coins by 100, we have

$3x = \frac{2}{7} \times 7x + 25$

$=> x = 25$

Thus, Suzy's coins

$= 4x = 4 \times 25 = 100.$

The correct answer is Option B.

87. Say Material cost, Labour cost, Factory overhead cost, and Office overhead cost are a, b, c, and d.

Since we have to deal with fractions, $\frac{3}{7}$, $\frac{4}{7}$, and $\frac{1}{2}$, let us assume that the total cost = $1,400.

Thus, we have...

$a + b = \frac{3}{7}$ of 1,400 = 600 ...(1)

$b + c = \frac{1}{2}$ of 1,400 = 700 ...(2)

$c + d = \frac{4}{7}$ of 1,400 = 800 ...(3)

$a + d = \frac{1}{2}$ of 1,400 = 700 ...(4)

Subtracting equation (2) from (1), we get

$c - a = 100 => c > a$; a is not the highest.

Subtracting equation (2) from (3), we get

$d - b = 100 => d > b$; b is not the highest.

We cannot establish whether $c >=< d$. Factory overhead cost or Office overhead cost can be highest.

The correct answer is Option E.

88. Let the number of points for the 1st question = x.

We know that each question is worth 2 points more than the preceding question.

Thus, the worth of each question in points for the 20 questions forms an arithmetic progression with the first term as x and a constant difference between consecutive terms of 2.

The nth term in arithmetic progression = $a + (n - 1) \times d$; (a is the first term, and d is the constant difference between consecutive terms)

Thus, the number of points for the 20th question = $x + (20 - 1) \times 2 = (x + 38)$.

Since the points for the above questions have a constant difference, the average points per question

$= \left(\frac{\text{First term} + \text{Last term}}{2}\right)$

$= \frac{x + (x + 38)}{2}$

$= (x + 19)$

Thus, the total points for all 20 questions $= 20 \times (x + 19)$.

Thus, we have

$20\,(x + 19) = 400$

$=> x = 1$

Thus, the number of points for the 4th question $= x + (4 - 1) \times 2 = 1 + 6 = 7$.

The correct answer is Option B.

89. Let the number of students in the school $= x$.

Thus, number of students taking the science course $= \left(80 + \frac{x}{3}\right)$.

Thus, the number of students taking chemistry $= \frac{1}{3}\left(80 + \frac{x}{3}\right)$.

Since $\left(\frac{1}{6}\right)^{th}$ of the students are taking chemistry, we have

$\frac{1}{3}\left(80 + \frac{x}{3}\right) = \frac{x}{6}$

$=> 80 + \frac{x}{3} = \frac{x}{2}$

$=> \frac{x}{2} - \frac{x}{3} = 80$

$=> \frac{x}{6} = 80$

$=> x = 480$

The correct answer is Option C.

90. Let the number of males = x.

Thus, the number of females = $(50 - x)$.

Thus, the number of employees who eat company breakfast

$$= \frac{x}{4} + \frac{50 - x}{5}$$

$$= \frac{5x + 4(50 - x)}{20}$$

$$= \frac{x + 200}{20}$$

$$= \frac{x}{20} + 10$$

Thus, to maximize the above value, we must have the largest possible value of x.

Also, since $\frac{x}{20}$ must be an integer, we have to choose $x = 40$.

Thus, the required maximum value = $\frac{40}{20} + 10 = 12$

The correct answer is Option D.

91. Let T, F and M be the total number of students, the number of female and the number of male, respectively.

We have

$$\frac{1}{8} \times F = \frac{1}{12} \times T$$

$$=> F = \frac{8}{12} \times T$$

$$=> F = \frac{2}{3} \times T$$

$$=> M = \left(1 - \frac{2}{3}\right) \times T$$

$$=> M = \frac{1}{3} \times T$$

$$=> \frac{M}{F} = \frac{\frac{1}{3} \times T}{\frac{2}{3} \times T}$$

$$=> \frac{M}{F} = \frac{\frac{1}{3}}{\frac{2}{3}} = \frac{1}{2}$$

Thus, the required ratio = $\frac{1}{2}$

The correct answer is Option C.

92. The incorrect total of the score of the class = $40 \times 32 = 1,280$.

The correct total of the score of the class = $1,280 - 30 + 40 = 1,290$.

The correct average (arithmetic mean) score of the class = $\frac{1,290}{40} = 32.25$.

The information that the average (arithmetic mean) score of the girls is 30 and that of boys is 40 is redundant.

The correct answer is Option C.

Alternate approach:

Since the correct score (40) is greater than the incorrect score (30), the average score would increase by $\frac{(40 - 30)}{40} = 0.25$.

Thus the correct average (arithmetic mean) score of the class = $32 + 0.25 = 32.25$.

93. We have 120 liters of 25% Chemical A solution.

Thus, amount of Chemical A = 25% of 120

$= \frac{25}{100} \times 120$

= 30 liters

Let x liters of Chemical A be added.

Thus, the final amount of Chemical A = $(x + 30)$ liters.

Total volume of the solution = $(x + 120)$ liters.

Since the final concentration of Chemical A is 40%, we have

$\left(\frac{x + 30}{x + 120}\right) \times 100 = 40$

=> $\left(\frac{x + 30}{x + 120}\right) = \frac{40}{100} = \frac{2}{5}$

=> $5x + 150 = 2x + 240$

=> $x = 30$

The correct answer is Option E.

Alternate approach:

Amount of water in the initial solution = $(100 - 25)\%$ of 120 = 75% of 120 = 90 liters.

Let x liters of Chemical A be added.

Total volume of the solution = $(x + 120)$ liters.

Final concentration of water = $(100 - 40)\% = 60\%$

Since water is not added, the quantity of water remains the same, i.e. 90 liters.

Thus, we have

$$\left(\frac{90}{x + 120}\right) \times 100 = 60$$

$$=> \frac{90}{x + 120} = \frac{60}{100} = \frac{3}{5}$$

$$=> 3x + 360 = 450$$

$$=> x = 30$$

94. Increase in the number of chickens from the first month (144) to the second month (c) $= (c - 144)$

Thus, fractional increase

$$= \frac{\text{Increase}}{\text{Original value}}$$

$$= \left(\frac{c - 144}{144}\right)$$

Increase in the number of chickens from the second month (c) to the third month (256) $= (256 - c)$

Thus, fractional increase

$$= \frac{\text{Increase}}{\text{Original value}}$$

$$= \left(\frac{256 - c}{c}\right)$$

Since the fractional increases are the same, we have

$$\frac{c - 144}{144} = \frac{256 - c}{c}$$

=> $\frac{c}{144} - 1 = \frac{256}{c} - 1$

=> $\frac{c}{144} = \frac{256}{c}$

=> $c^2 = 144 \times 256$

=> $c = 12 \times 16$

=> $c = 192$

The correct answer is Option A.

Alternate approach 1:

Since the number of chickens increased by the same fraction during each of the two periods, we can say that the ratio of the number of chickens in consecutive intervals would be equal.

Thus we have

$\frac{144}{c} = \frac{c}{256}$

=> $c^2 = 144 \times 256$

=> $c = 12 \times 16$

=> $c = 192$

Alternate approach 2:

Since the fractional increase in each period is the same, we can conclude that the percent change in each period is also the same.

Let the percent change be r.

Thus, using the concept of compound interest, we have

$256 = 144 \times \left(1 + \frac{r}{100}\right)^2$

=> $\left(1 + \frac{r}{100}\right)^2 = \left(\frac{16}{12}\right)^2$

=> $1 + \frac{r}{100} = \frac{16}{12}$

=> $1 + \frac{r}{100} = \frac{4}{3}$

=> $\frac{r}{100} = \frac{1}{3}$

$=> r = \frac{100}{3}\%$

$=>$The number of chickens by second month, $c = 144 \times \left(1 + \frac{100/3}{100}\right)^1 = 144 \times \left(\frac{4}{3}\right) = 192$

4.6 Speed, Time & Distance

95. Let the average speed for the trip be S.

T_1 = Time required to travel 900 miles with speed $S = \frac{900}{S}$ hours

S is increased by 10 miles; so new speed = $(S + 10)$ mph

T_2 = Time required to travel 900 miles with speed $(S + 10) = \frac{900}{S + 10}$ hours

Given that the difference between T_1 and T_2 is 1 hour

$$=> \frac{900}{S} - \frac{900}{S + 10} = 1$$

$$=> 900\left(\frac{1}{S} - \frac{1}{S + 10}\right) = 1$$

$$=> \frac{S + 10 - S}{S(S + 10)} = \frac{1}{900}$$

$$=> 10 \times 900 = S^2 + 10S$$

$$=> S^2 + 10S - 9,000 = 0$$

Roots of above quadratic equations are '90' and '−100'.

Since speed cannot be negative, '−100' is ignored, so $S = 90$ mph.

The correct answer is Option E.

96. The truck traveled 4 miles less per gallon on the state highway compared to on the national highway.

Let's consider if the truck travels x miles per gallon on the national highway then the truck travels $(x - 4)$ miles per gallon on the state highway.

Capacity of the full tank of diesel on the national highway = $\frac{336}{x}$ gallons

Capacity of the full tank of diesel on the state highway = $\frac{224}{x - 4}$ gallons

Since the above fractions must be equal, we have

$$\frac{336}{x} = \frac{224}{x - 4}$$

$$=> \frac{3}{x} = \frac{2}{x - 4}$$

$$=> 3x - 12 = 2x$$

$=> x = 12$

As the truck travels $(x - 4)$ miles per gallon on the state highway, required answer is $12 - 4 = 8$.

The correct answer is Option B.

97. Time required to travel 10 miles at speed 60 miles per hour

$= \frac{10}{60}$ hours $= \frac{10}{60} \times 60$ minutes $= 10$ minutes

Now new time required $= 10 + 5 = 15$ minutes $= \frac{15}{60} = \frac{1}{4}$ hours.

$$\text{Speed} = \frac{\text{Distance}}{\text{Time}}$$

$$= \frac{10 \text{ miles}}{15 \text{ Minutes}} = \frac{10}{\frac{1}{4}} = 40 \text{ miles per hour}$$

The correct answer is Option B.

98. Speed for first 10-minutes interval = 30 miles per hour

Speed for second 10-minutes interval = 40 miles per hour

Speed for third 10-minute interval = 50 miles per hour

Speed for fourth 10-minute interval = 60 miles per hour

Distance travelled in the fourth 10-minute interval $= \frac{60}{60} \times 10 = 10$ miles

The correct answer is Option D.

99. Time taken for the onward journey $= \frac{600}{500} = \frac{6}{5}$ hours.

Time taken for the return journey $= \frac{600}{400} = \frac{3}{2}$ hours.

Thus, total time taken for the round trip $= \frac{6}{5} + \frac{3}{2} = \frac{27}{10}$ hours.

Total distance travelled $= 2 \times 600 = 1200$ miles.

Thus, average speed $= \frac{\text{Total distance}}{\text{Total time}} = \frac{1,200}{\frac{27}{10}} = 1200 \times \frac{10}{27} = 1,000 \times \frac{4}{9} =\approx 444$ miles/hr.

The correct answer is Option B.

Alternate approach:

Since the distance travelled for the onward and the return journey is the same, we have

$$\text{Average speed} = \frac{2 \times \text{Speed}_1 \times \text{Speed}_2}{\text{Speed}_1 + \text{Speed}_2} = \frac{2 \times 500 \times 400}{500 + 400} = \frac{2 \times 500 \times 400}{900} = \frac{4,000}{9} =\approx 444 \text{ mph}$$

100. Average speed of the truck for 1/2 of 800 miles (= 400 miles) = 40 miles per hour

Thus, time taken to cover the above distance (400 miles) = $\frac{400}{40}$ = 10 hours

Average speed of the truck or the entire 800 miles = 50 miles per hour

Thus, time taken to cover the total distance = $\frac{800}{50}$ = 16 hours

Thus, time taken by the truck to cover the remaining miles 400 miles = 16 – 10 = 6 hours

Thus, average speed of the truck for the remaining miles 400 miles = $\frac{400}{6}$ = 66.66 =≈ 67 miles per hour.

The correct answer is Option D.

101. Let the marathoner's speed on the first day be x miles per hour.

Thus, the his speed on the second day was $(x + 3)$ miles per hour.

Let the time for which the marathoner ran on the first day be t hours.

Thus, the time for which he ran on the second day = $(8 - t)$ hours.

Distance covered by him on the first day = $(x \times t)$ miles.

Distance covered by him on the second day = $((x + 3) \times (8 - t))$ miles.

Thus, total distance covered in two days = $[xt + (x + 3)(8 - t)]$ miles.

Thus, we have

$xt + (x + 3)(8 - t) = 36$

$\Rightarrow xt + 8x - xt + 24 - 3t = 36$

$\Rightarrow 8x = 3t + 12$

Since this is a single linear equation with two variables, we cannot get the unique value of x; however, we can get consistent values of x.

Working with the options one at a time:

Option A: $x = 0.5 \Rightarrow t = -4$ – Not possible, since t cannot be negative

We can deduce that since this is a 'Could be value' type of question, and we see that the smallest option value of x yields negative value for t, thus we should first try the largest value of x among the options.

Option E: $x = 2 \Rightarrow t = 4/3$ - Possible

Since this is a 'Could be value' type of question, and only one option is correct, Option E is the correct answer. There is no need to check other options since two or more options cannot be simultaneously correct.

Thus, we have $x = 2$

The correct answer is Option E.

Alternate approach:

Average speed of the marathoner for the two days = $\left(\frac{\text{Total distance}}{\text{Total time}}\right) = \frac{36}{8} = 4.5$ miles per hour.

If the average speed on the first day be x miles per hour, the average speed on the second day should be $(x + 3)$ miles per hour.

Thus, the average speed of 4.5 miles per hour must lie between the values of the speeds on the two days.

Thus, we have

$x < 4.5 < x + 3$

Among the options, only $x = 2$ satisfies.

102. On seeing this question, one would immediately calculate the time taken to meet using the data given.

From there on, one would try to find the distance between the trains 2 hours before they meet.

However, such calculations are not necessary.

The question simply asks us, "If two trains traveling at 50 miles per hour and 60 miles per hour need 2 hours to meet, how far away are they from one another."

Since both trains travel for 2 hours before they meet, one train travels $50 \times 2 = 100$ miles and the other train travels $60 \times 2 = 120$ miles.

Thus, together the trains travel $(100 + 120) = 220$ miles before they meet.

Thus, distance between the trains 2 hours before they meet = 220 miles.

The correct answer is Option E.

103. Length of section = 10 miles.

Original speed limit = 50 miles per hour.

Thus, time taken to cover this distance $= \frac{10}{50} = \frac{1}{5}$ hours

New speed limit = 40 miles per hour.

Thus, time taken to cover this distance $= \frac{10}{40} = \frac{1}{4}$ hours

Thus, the required difference between the time durations is given $= \left(\frac{1}{4} - \frac{1}{5}\right)$ hours

$= \frac{1}{20}$ hours

$= \frac{1}{20} \times 60 = 3$ minutes

The correct answer is Option A.

104. We know that

Speed of Train B = 25% more than the speed of Train A

=> Speed of Train B = (100 + 25)% of the speed of Train A

=> Speed of Train B = 125% of the speed of Train A

=> Speed of Train B $= \frac{125}{100} \times$ (Speed of Train A)

=> Speed of Train B $= \frac{5}{4} \times$ (Speed of Train A)

=> $\frac{\text{Speed of Train B}}{\text{Speed of Train A}} = \frac{5}{4}$

Since time is inversely proportional to speed for a constant distance, we have

$$\frac{\text{Time taken by Train B to cover a distance}}{\text{Time taken by Train A to cover the same distance}} = \frac{1}{\frac{5}{4}} = \frac{4}{5}$$

We know that Train A took 4 hours to cover the distance.

Thus, we have

=> Time taken by Train B to cover a distance $= 4 \times \frac{4}{5} = \frac{16}{5} = 3\frac{1}{5}$ hours

The correct answer is Option C.

Alternate approach:

Train A took 4 hours to travel 100 miles.

Thus, the average speed of Train A = 25 miles per hour.

Since the average speed of Train B is 25% more than that of Train A, we have

Average speed of Train B = 25 + 25% of 25 = $\frac{125}{4}$ miles per hour.

Thus, time required to travel 100 miles by Train B

$= \frac{\text{Distance}}{\text{Speed}} = \frac{100}{\frac{125}{4}} = \frac{16}{5} = 3\frac{1}{5}$ hours

105. Jeff's speed = 40 miles per hour.

Thus, distance covered by Jeff in 60 minutes = 40 miles.

Thus, distance covered by Jeff in 36 minutes $= \frac{40}{60} \times 36 = 24$ miles.

This distance of 24 miles is 3 times of what Amy drives in 30 minutes.

Thus, distance covered by Amy in 30 minutes $= \frac{24}{3} = 8$ miles.

Thus, distance covered by Amy in 60 minutes $= \frac{60}{30} \times 8 = 16$ miles.

Thus, Amy's speed = 16 miles per hour.

The correct answer is Option C.

106. Time taken for the first bus to cover the distance between the two depots A & B = 5 hours.

Time taken for the second bus to cover the distance between the two depots A & B = 3 hours.

Since we need to find the time when the two buses pass one another, the actual length of the distance is not required.

So, we can assume a suitable value of the distance for ease of calculations.

Let the distance between the depots = least common multiple of 5 and 3 = 15 miles.

Thus, speed of the first bus = $\frac{15}{5}$ = 3 miles/hour.

Speed of the second bus = $\frac{15}{3}$ = 5 miles/hour.

We know that the first bus started one hour earlier than when the second bus started.

Thus, in 1 hour, the distance covered by the first bus = $3 \times 1 = 3$ miles.

Thus, at 8:00 am, the distance between the two buses = $(15 - 3) = 12$ miles.

At 8:00 am, both buses approach one another at speeds 3 miles/hour and 5 miles/hour.

Thus, when the buses pass one another, the ratio of the distances covered by the trains would be equal to the ratio of their respective speeds = 3 : 5

Thus, distance covered by the first bus = $\left(\frac{3}{3+5}\right) \times 12 = 4.5$ miles.

Time taken by the first bus to cover 4.5 miles = $\frac{4.5}{3}$ = 1.5 hours = 1 hour 30 minutes.

The buses will meet at 8:00 am + 1 hour 30 minutes = 9:30 am.

Alternately, we can use the concept of relative speed:

Since the buses travel in opposite directions, their relative speed = $(3 + 5) = 8$ miles/hour.

Thus, time taken to cover 12 miles = $\frac{12}{8}$ = 1.5 hours = 1 hour 30 minutes.

Thus, the time when the two buses pass one another = 1 hour 30 minutes past 8:00 am

The correct answer is Option C.

4.7 Time & Work

107. Number of copies made per hour by A = 1,500.

Thus, number of copies made per hour by B = $2 \times 1,500 = 3,000$
(Since it is twice as efficient as the first machine)

Since B works 12 hours a day, number of copies made in a day

= $3,000 \times 12$ = 36,000

Thus, number of copies made in 20 days = $36,000 \times 20$ = 720,000

The correct answer is Option E.

108. According to the given data, the pump filled $\left(\frac{5}{6} - \frac{1}{2}\right) = \frac{1}{3}$ part of the pool in $2\frac{1}{3}\left(= \frac{7}{3}\right)$ hours

Thus, time taken by the pump to fill the entire pool = $\frac{\left(\frac{7}{3}\right)}{\left(\frac{1}{3}\right)} = \frac{7}{3} \times \frac{3}{1} = 7$ hours

The correct answer is Option C.

109. Let the volume of the pool = LCM (10, 15) = 30 liters

The first pump can fill the pool in 10 hours.
Thus, the rate of the first pump = $\frac{30}{10}$ = 3 liters per hour.

The second pump can fill the pool in 15 hours.
Thus, the rate of the second pump = $\frac{30}{15}$ = 2 liters per hour.

The first pump was on for the entire 7 hours in which it filled = 7×3 = 21 liters.

Thus, the remaining (30 – 21) = 9 liters were filled by the second pump.

Time taken for the second pump to fill 9 liters = $\frac{9}{2}$ = 4.50 hours.

Thus, of the total 7 hours, the second pump was on for 4.50 hours.

Thus, the first pump alone was on for (7 – 4.50) = 2.50 hours.

Thus, h = 2.50

The correct answer is Option B.

110. The emptying pipe can empty the pool which is $\left(\frac{3}{4}\right)^{\text{th}}$ full in 9 hours.

Thus, time taken to empty the entire pool = $9 \times \frac{4}{3} = 12$ hours.

It is given that capacity of swimming pool is 5,760 gallons.

Thus, the rate at which the emptying pipe removes water
$= \frac{5,760}{12} = 480$ gallons per hour.

The rate at which the pool can be filled
= 12 gallons per minute

$= 12 \times 60 = 720$ gallons per hour.

Thus, the effective filling rate when both filling and emptying occur simultaneously
$= 720 - 480 = 240$ gallons per hour.

Since we need to fill only half the pool, the volume required to be filled
$= \frac{5,760}{2} = 2,880$ gallons.

Thus, time required $= \frac{2,880}{240} = 12$ hours.

The correct answer is Option B.

111. We know that lathe machine B manufactures 300 X-type bearings in 60 days.

Since lathe machine A manufactures bearings thrice as fast as machine B does, time taken by lathe machine A to manufacturer 300 X-type bearings
$= \frac{60}{3} = 20$ days

Since each Y-type bearing takes 2.5 times the time taken to manufacturer each X-type bearing, the time taken by lathe machine A to manufacturer 300 Y-type bearings
$= 20 \times 2.5 = 50$ days

Thus, the number of Y-type bearings manufactured by lathe machine A in 50 days = 300.

Thus, the number of bearings manufactured by lathe machine A in 10 days

$= \frac{10}{50} \times 300 = 60$

The correct answer is Option D.

112. Numbers of copies made by Photocopier A in 1 hour = $\frac{1,200}{3} = 400$

Numbers of copies made by Photocopier B in 1 hour = $\frac{1,200}{2} = 600$

Numbers of copies made by Photocopier C in 1 hour = $\frac{1,200}{6} = 200$

Thus, when all the three photocopiers work together, total numbers of copies made in 1 hour

$= 400 + 600 + 200 = 1,200$ copies

Thus, time taken to make 3,600 copies = $\frac{3,600}{1,200} = 3$ hours.

The correct answer is Option D.

113. Cost of food consumed by 5 men in 4 days = \$150

=> Cost of food consumed by 1 man in 4 days = $\$\left(\frac{150}{5}\right) = \$30$

=> Cost of food consumed by 1 man in 1 day = $\$\left(\frac{30}{4}\right) = \$\left(\frac{15}{2}\right)$

We know that one woman consumes three-fourth the amount of food consumed by a man.

Thus, 2 women consume food equivalent to $2 \times \left(\frac{3}{4}\right) = \frac{3}{2}$ men.

Thus, 4 men and 2 women are equivalent to $\left(4 + \frac{3}{2}\right) = \frac{11}{2}$ men.

Thus, cost of food consumed by $\frac{11}{2}$ men in 1 day = $\$\left(\frac{15}{2} \times \frac{11}{2}\right) = \$\left(\frac{165}{4}\right)$

=> Cost of food consumed by $\frac{11}{2}$ men in 8 days = $\$\left(\frac{165}{4} \times 8\right) = \$330$

The correct answer is Option B.

114. Let the time taken by Mark and Kate working together = x hours

Thus, time taken by Mark, working alone = $(x + 12)$ hours

Time taken by Kate, working alone = $(x + 27)$ hours

Let us assume the total work to be 1 unit.

Part of work done by Mark and Kate, working together, in 1 hour = $\left(\frac{1}{x}\right)$

Part of work done by Mark in 1 hour = $\left(\frac{1}{x + 12}\right)$

Part of work done by Kate in 1 hour = $\left(\frac{1}{x+27}\right)$

Thus, we have

$$\frac{1}{x} = \frac{1}{x+12} + \frac{1}{x+27}$$

$$=> \frac{1}{x} = \frac{(x+27)+(x+12)}{(x+12)(x+27)}$$

$$=> (x+12)(x+27) = x(2x+39)$$

$$=> x^2 + 39x + 324 = 2x^2 + 39x$$

$$=> x^2 = 324$$

Since x is positive, we have

$$x = \sqrt{324} = 18$$

The correct answer is Option C.

Alternate approach:

Work done by Mark and Kate in x hours = Work done by Mark in $(x + 12)$ hours

Cancelling work done by Mark in x hours from both sides:

Work done by Kate in x hours = Work done by Mark in 12 hours

=> Work done by Kate in 1 hour = Work done by Mark in $\left(\frac{12}{x}\right)$ hours . . . (i)

Again, we have

Work done by Mark and Kate in x hours = Work done by Kate in $(x + 27)$ hours

Cancelling work done by Kate in x hours from both sides:

Work done by Mark in x hours = Work done by Kate in 27 hours

=> Work done by Kate in 27 hours = Work done by Mark in x hours

=> Work done by Kate in 1 hour = Work done by Mark in $\left(\frac{x}{27}\right)$ hours . . . (ii)

Thus, from (i) and (ii), we have

$$\frac{12}{x} = \frac{x}{27}$$

=> $x^2 = 324$

=> $x = 18$

4.8 Computational

115. Amount of chemical evaporated in y minutes = x liters.

Thus, the amount of chemical evaporated in 1 minute = $\frac{x}{y}$ liters.

Thus, the amount of chemical evaporated in z minutes = $\frac{xz}{y}$ liters.

Cost of 1 liter of the chemical = \$25.

Thus, cost of the chemical evaporated in z minutes = $\$\left(25 \times \frac{xz}{y}\right) = \$\left(\frac{25xz}{y}\right)$.

The correct answer is Option D.

116. We know that the total cost of producing 25,000 pens is \$37,500 and the total cost of producing 35,000 pens is \$47,500.

Since it is given that total cost of producing pens is governed by a linear function, total cost must have a fixed cost component and a variable complement.

Note that fixed cost component is same, irrespective of how many pens Company X makes.

Thus, incremental cost (variable complement) of producing additional 10,000 pens (= 35,000 – 25,000) pens = \$47,500 – \$37,500 = \$10,000.

=> Thus, incremental cost (variable complement) of producing additional 1 pen = $\frac{10,000}{10,000} = \$1$

Thus, incremental cost (variable complement) of producing additional 15,000 pens (= 50,000 – 35,000) pens = $1 \times 15,000 = \$15,000$

=> Thus, total cost of producing 50,000 pen = total cost of producing 35,000 pen + incremental cost (variable complement) of producing additional 15,000 pens = \$47,500 + \$15,000 = \$62,500

The correct answer is Option C.

117. The cost of four pencils = $\$\,(1.35 \times 4) = \$5.40$

The cost of two erasers = $\$\,(0.30 \times 2) = \$0.60$

Thus, total cost = $\$(5.40 + 0.60) = \$6.00$

Thus, Suzy has one-third of the above amount = $\$\left(\frac{6}{3}\right) = \$2$

The correct answer is Option B.

118. Total increase in population = 378 – 360 = 18 million

Increase in population per month = 30,000

Thus, increase in population per year = $30,000 \times 12$ = 360,000 = 0.36 million

Thus, number of years required for the increase = $\frac{18}{0.36}$ = 50 years

Thus, the population would be 378 million in the year (2012 + 50) = 2062

The correct answer is Option C.

119. The restaurant uses $\frac{1}{2}$ cup milk-cream in each serving of its ice-cream.

Since each carton has $2\frac{1}{2} = \frac{5}{2}$ cups of milk-cream, number of servings of ice-cream possible using one carton $= \frac{\left(\frac{5}{2}\right)}{\left(\frac{1}{2}\right)} = 5$

Thus, number of cartons required for 98 servings of the ice-cream = $\frac{98}{5}$ =≈ 19.3

However, the number of cartons must be an integer.

Thus, the minimum number of cans required is 20.

The correct answer is Option C.

120. We need to minimize the total number of coins such that each box has at least 2 coins.

We know that at the most 3 boxes can have the same number of coins.

Since we need to minimize the total number of coins, we must have as many boxes having the same number (minimum possible number, i.e. 2 coins) of coins as possible.

Thus, for each of the 3 boxes containing an equal number of coins, we have '2' coins.

Thus, number of coins in the 3 boxes = $2 \times 3 = 6$.

Since each of the remaining 4 boxes have a different number of coins, let us put in 3, 4, 5, and 6 coins in those boxes.

Thus, the total number of coins = 6 + (3 + 4 + 5 + 6) = 24.

The correct answer is Option C.

121. Distance covered by the lava in 1 hour = $\frac{15}{4}$ feet

$= \frac{\frac{15}{4}}{5,280}$ miles $= \frac{15}{4 \times 5,280}$ miles

Hours needed to cover $\frac{3}{2}$ miles $= \frac{\frac{3}{2}}{\frac{15}{4 \times 5,280}}$

$= \frac{3 \times 4 \times 5,280}{2 \times 15} = 2,112$ hours

Days needed to cover $\frac{3}{2}$ miles $= \frac{2,112}{24} = 88$ days

The correct answer is Option E.

122. Cross-sectional area of the bar = 4 square feet

Rate at which the bar moves through the conveyor

= 360 feet per hour

$= \frac{360}{60 \times 60}$ feet per second

= 0.1 feet per second

Thus, volume of the bar that moves through the conveyor per second

$= 0.1 \times 4$ cubic feet

= 0.4 cubic feet

Thus, time taken to move 0.4 cubic feet of bar through the conveyor = 1 second.

Thus, time taken to move 8.4 cubic feet of bar through the conveyor $= \frac{8.4}{0.4} = 21$ seconds.

The correct answer is Option A.

123. We know that for a salary grade G, the hourly wage W, in dollars, is given by:

$W = 1,140 + 45(G - 1)$

Thus, for a salary grade $G = 1$, the corresponding hourly wage W

$= \$\,(1,140 + 45\,(1 - 1))$

$= \$1,140$

Again, for a salary grade $G = 7$, the corresponding hourly wage W

$= \$ (1,140 + 45(7 - 1))$

$= \$(1,140 + 270)$

$= \$1,410$

Thus, the required difference $= \$ (1,410 - 1,140) = \$270$.

The correct answer is Option B.

Alternate approach:

We have wage equation: $W = 1,140 + 45(G - 1)$

$=> W = 1,140 - 45 + 45G$

We see that in the above equation, $45G$ is variable and '$1,140 - 45$' is a constant for the workers of all the grades.

Thus, the difference between the remuneration would result because of $45G$.

Thus, difference in remuneration of grade-7 and grade-1 worker $= 45 \times 7 - 45 \times 1 = \$270$.

124. Considering the radios sold:

We know that the price of the certain radio was the 15th highest price as well as the 20th lowest price among the prices of the radios sold.

Thus, the number of radios sold at a price greater than the price of the particular radio $= 15 - 1 = 14$

Also, the number of radios sold at a price lower than the price of the particular radio $= 20 - 1 = 19$

Thus, total number of radios sold $= 14 + 19 + 1$ (including that particular radio) $= 34$

Considering all the items sold:

We know that the price of the certain DVD player was the 29th highest price as well as the 37th lowest price among the prices of all the items sold.

Thus, the number of items sold at a price greater than the price of the particular item (i.e. DVD player) $= 29 - 1 = 28$

Also, the number of items sold at a price lower than the price of the particular item (i.e. DVD player) = 37 – 1 = 36.

Thus, total number of items sold = 28 + 36 + 1 (including that particular radio) = 65

Thus, the number of DVD players = Total number of items – number of radios = 65 – 34 = 31

The correct answer is Option B.

125. Commission received on the first 150 orders at the rate of $25 per order

$= \$\,(25 \times 150) = \$3,750$

Total commission received received = $5,000.

Thus, commission received on the additional orders (above 150) at $12.50 per order

$= \$(5,000 - 3,750) = \$1,250$

Thus, number of additional order made= $\frac{1,250}{12.5}$

= 100

Thus, total number of order = Initial 100 orders + Additional 150 orders

= 250 orders

The correct answer is Option D.

126. The loss would be minimum if fewer number of $50 checks were lost as compared to $20 checks.

Thus, the traveler should have cashed a greater number of $50 checks than $20 checks.

We know that the number of $20 checks cashed was 2 more or 2 less than the number of $50 checks cashed.

Thus, from the above reasoning, we can conclude that the number of $20 checks was 2 less than the number of $50 checks.

Since the total number of checks cashed was 10, from the above information, we have

Number of $50 checks cashed = 6

Number of $20 checks cashed = 4

Alternately:

Let the number of $50 checks cashed = x

Thus, the number of \$20 checks cashed = $(x - 2)$

Since the total number of checks cashed is 10, we have

$x + (x - 2) = 10$

$\Rightarrow x = 6$

Thus, the total value of \$20 checks cashed = \$ $(4 \times 20) = \$80$.

The total value of \$50 checks cashed = \$ $(6 \times 50) = \$300$.

Thus, the total value of all the checks cashed = \$ $(80 + 300) = \$380$.

We know that the total value of all the checks with him was \$2,000.

Thus, the total value of all the checks lost = \$ $(2,000 - 380) = \$1,620$.

The correct answer is Option D.

127. Let's list down per kg prices realized upon buying three pack-sizes.

1. 5-kg pack for \$16 => $\frac{16}{5}$ = \$3.20 per kg.
2. 10-kg pack for \$26 => $\frac{26}{10}$ = \$2.60 per kg.
3. 25-kg pack for \$55 => $\frac{55}{25}$ = \$2.20 per kg.

From the above calculation, it is obvious that larger the pack size, smaller the per kg amount a customer has to pay.

Let's take few options the customer can take to buy 40 kg dog food.

1. Buy 50 kg.: 2 packs of 25 kg

Cost = $2 \times 55 = \$110$; though the quantity of food (50 kg) is more than the minimum required (40 kg), we must take this option into consideration since the per kg price of the food is least for a 25 kg pack.

2. Buy 45 kg.: 1 pack of 25 kg and 2 packs of 10 kg

Cost = $55 + 2 \times 26 = \$107$

3. Buy 40 kg.: 1 pack of 25 kg, 1 pack of 10 kg, and 1 pack of 5 kg

Cost = $55 + 26 + 16 = \$97$

From the above, it is clear that the customer must buy 40 kg (1 pack of 25 kg, 1 pack of 10 kg, and 1 pack of 5 kg), paying the minimum price of \$97.

The correct answer is Option C.

4.9 Interest

128. Amount after 4 years = \$3,200
Amount after 6 years = \$3,800

Thus, interest accumulated in 2 (= 6 – 4) years = $\$\,(3,800 - 2,200) = \$600$

Thus, interest accumulated per year $= \$\left(\frac{600}{2}\right) = \$300$ (since under simple interest, interest accumulated every year is constant)

Thus, interest accumulated in the first 4 years $= \$\,(300 \times 4) = \$1,200$

Thus, principal amount invested $= \$\,(3,200 - 1,200) = \$2,000$

Thus, on \$2,000 invested, interest accumulated is \$300 every year.

Thus, rate of interest $= \frac{300}{2,000} \times 100 = 15\%$

The correct answer is Option C.

129. Simple interest accumulated after 2 years = \$600

Thus, simple interest per year $= \$\left(\frac{600}{2}\right) = \$300$ (since under simple interest, interest accumulated every year is constant)

Thus, compound interest accumulated after the first year = \$300 (equal to the simple interest accumulated after one year)

Thus, compound interest accumulated in the second year $= \$\,(300 + 63) = \$363$ (since the total compound interest accumulated in 2 years is \$63 more than that under simple interest)

The higher interest in the second year is due to the additional interest on the interest accumulated after one year.

Thus, we can say that interest on \$300 in one year = \$63

Thus rate of interest $= \frac{63}{300} \times 100 = 21\%$

The correct answer is Option D.

130. Let the sums borrowed at 10% and 8% rate of interest be $\$x$ each.

Let the time after which Suzy repays the second sum be t years.

Thus, the time after which she repays the first sum is $(t - 1)$ years.

Since the amount to be repaid in either case is the same, the interest accumulated is also equal.

Simple Interest $= \left(\frac{PRT}{100}\right)$, where P =Principal, R = Rate of Interest, and T = Time Interval

Hence, we have

$$\frac{10 \times x \times (t-1)}{100} = \frac{8 \times x \times t}{100}$$

$=> 5(t-1) = 4t$

$=> t = 5$ years

Thus, the time for which she borrowed the first sum of money $= t - 1 = 4$ years

The correct answer is Option D.

131. Total amount invested = $100,000.

Let $x be invested at 3% and $(100,000 − x) be invested at 4%

Thus, at the end of 1 year, interest on $x

$$= \frac{x \times 3 \times 1}{100} = \frac{3x}{100}$$

Also, at the end of 1 year, interest on $(100,000 − x)

$$= \frac{(100,000 - x) \times 4 \times 1}{100} = \frac{4\,(100,000 - x)}{100}$$

Since the total interest is $3,600, we have

$$\frac{3x}{100} + \frac{4\,(100,000 - x)}{100} = 3,600$$

$=> 400,000 - x = 360,000$

$=> x = 40,000$

$=> 100,000 - x = 60,000$

Thus, the fraction of the total invested at 4% $= \frac{60,000}{100,000} = \frac{3}{5}$

The correct answer is Option D.

Alternate approach:

We can use the method of alligation:

Let the amounts invested at 3% and 4% be $\$x$ and $\$y$, respectively.

The total interest on \$100,000 is \$3,600.

Thus, the effective rate of interest as a whole $= \frac{3,600}{100,000} \times 100 = 3.6\%$.

Thus, we have

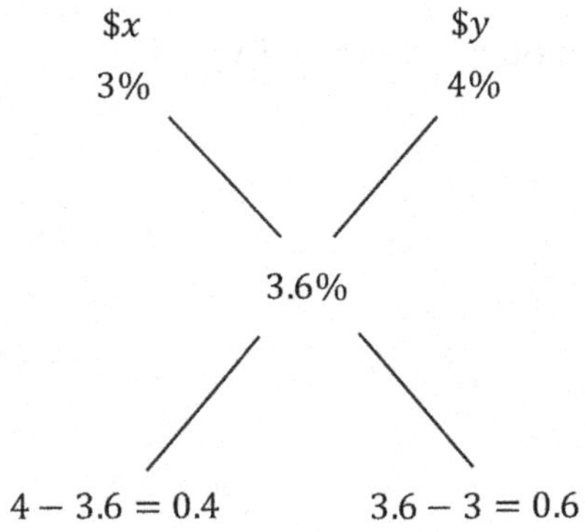

Hence, we have

$$\frac{x}{y} = \frac{0.4}{0.6} = \frac{2}{3}$$

Hence, the required fraction $= \frac{3}{2+3} = \frac{3}{5}$

132. Let the sum of money invested at 4% rate of interest be $\$x$.

Thus, the sum of money invested at 7% rate of interest $= \$\,(18,000 - x)$

Thus, interest on $\$x$ at 4% rate of interest in 2 years

$$= \$\left(\frac{x \times 2 \times 4}{100}\right) = \$\left(\frac{8x}{100}\right)$$

Also, interest on $\$\,(18,000 - x)$ at 7% rate of interest in 2 years

$$= \$\left(\frac{(18,000 - x) \times 2 \times 7}{100}\right) = \$\left(\frac{14\,(18,000 - x)}{100}\right)$$

Since total interest is \$2,100, we have

$$\frac{8x}{100} + \frac{14\,(18,000 - x)}{100} = 2,100$$

$=> 14 \times 18,000 - 6x = 210,000$

$=> 6x = 252,000 - 210,000$

$=> x = 7,000$

The correct answer is Option C.

133. Let the sums of money invested at 10% and 20% rates of interest be $\$x$ each.

Since the difference between the interests earned after three years is between \$120 and \$140, we have

$$120 < \left(\frac{x \times 20 \times 2}{100}\right) - \left(\frac{x \times 10 \times 2}{100}\right) < 140$$

$$=> 120 < \frac{20x}{100} < 140$$

$=> 600 < x < 700$

We need to find the difference between the amounts earned after 2 years at compound interest at the same rates as above.

Thus, the required difference

$$= x\left(1 + \frac{20}{100}\right)^2 - x\left(1 + \frac{10}{100}\right)^2 = x\left\{\left(1 + \frac{20}{100}\right)^2 - \left(1 + \frac{10}{100}\right)^2\right\}$$

$$= x\left\{\left(1 + \frac{20}{100}\right) - \left(1 + \frac{10}{100}\right)\right\}\left\{\left(1 + \frac{20}{100}\right) + \left(1 + \frac{10}{100}\right)\right\}$$

$$= x\left(\frac{10}{100}\right)\left(2 + \frac{30}{100}\right) = x\,(0.1 \times 2.3) = x \times 0.23$$

We know that $600 < x < 700$

Thus, the required difference $(x \times 0.23)$ is

$600 \times 0.23 < x \times 0.23 < 700 \times 0.23$

$=> 138 <$ Required difference < 161

Thus, the required difference lies between \$138 and \$161.

Only Option D lies in the above range.

The correct answer is Option D.

134. Population at the start of the experiment = x

Increase in population at the end of the 1st month = $2x$

Thus here we can say that, rate of increase = $\left(\frac{2x}{x}\right) \times 100 = 200\%$

This 200% increase remains same for each of the next 4 months.

Thus, applying the concept of compounding, we have

$$x\left(1 + \frac{200}{100}\right)^5 > 1,000$$

$$=> x > \frac{1,000}{(3)^5}$$

$$=> x > \frac{1,000}{243}$$

$$=> x > \left[\frac{1,000}{=\approx 250} = \approx 4\right]$$

Since we $250 > 243$, thus $\frac{1,000}{243} > 4 => x > 4$

Since x must be an integer value (it represents the number of organisms), the minimum possible value of $x = 5$.

The correct answer is Option D.

Alternate approach:

Population at the start of the experiment = x.

Increase in population at the end of the 1st month = $2x$.

Thus, population size at the end of the 1st month = $x + 2x = 3x$.

Increase in population after the 2nd month = $2 \times 3x = 6x$.

Thus, population size at the end of the 2nd month = $3x + 6x = 9x$.

Thus, we observe that the population size triples after every month.

Thus, the population size at the end of the 5th month

= $3 \times$ (The population size at the end of the 4th month)

= $3 \times 3 \times$ (The population size at the end of the 3rd month)

$= 3 \times 3 \times 3 \times$ (The population size at the end of the 2nd month) $= 3^3 \times 9x$

$= 243x$

Thus, we have

$243x > 1,000$

$=> x > \frac{1,000}{243}$

$=> x > 4.1$

Since x must be an integer value (it represents the number of organisms), the minimum possible value of $x = 5$.

135. The description about the calculation of interest is basically follows the concept of compound interest.

Since interest is calculated after every two-month period, in a year, it will be calculated 6 times. Also the rate on interest given is 12% per annum, thus, the rate of interest per period would be $12/6 = 2\%$ per two-month period.

The amount (A) under compound interest on a sum of money (P) invested at ($r\%$) rate of interest for n periods is given by:

$$A = P\left(1 + \frac{r}{100}\right)^n$$

We have to find out the value of $\left(\frac{A}{P}\right)$.

We know that $n = 6$ two-month periods and $r = 2\%$ per two-month period

Thus, $\frac{A}{P} = \left(1 + \frac{2}{100}\right)^6$

$\frac{A}{P} = (1.02)^6$

The correct answer is Option C.

136. The first $\$x$ deposited in the account earned interest for 2 years, while the additional $\$x$ earned interest for only 1 year.

The amount A under compound interest on a sum of money P invested at $r\%$ rate of interest for t years is given by:

$$A = P\left(1 + \frac{r}{100}\right)^t$$

Thus, the final value after 2 years for the first $\$x$ deposited

$= \$x\left(1 + \frac{4}{100}\right)^2$

$= \$(x \times 1.04^2)$

The final value after 1 year of the additional $\$x$ deposited

$= \$x\left(1 + \frac{4}{100}\right)^1$

$= \$1.04x$

Thus, total value of the money present in the account (y)

$= \$\left\{\left(x \times 1.04^2\right) + (x \times 1.04)\right\}$

$= \$\left(x\left(1.04^2 + 1.04\right)\right)$

Thus, we have

$y = x\left(1.04^2 + 1.04\right)$

$x = \frac{y}{\left(1.04^2 + 1.04\right)}$

$x = \frac{y}{1.04\,(1.04 + 1)}$

$x = \frac{y}{1.04 \times 2.04}$

The correct answer is Option D.

137. Let us understand the concept of the interest in the nth year with an example:

Let \$100 be invested at 10% rate of interest.

Interest accumulated after one year (or in the **first year**)

= \$ (10% of 100) = \$10

Thus, amount after one year = \$ (100 + 10) = \$110

Thus, interest in the **second year**

= \$ (10% of 110) = \$11 (= 110% of interest in the first year)

Thus, total interest accumulated in two years

= (Interest in the first year + Interest in the second year)

= \$ (10 + 11) = \$21

Thus, amount after two years = \$ (100 + 21) = \$121

Thus, interest in the **third year**

\$ (10% of 121) = \$12.10 (= 110% of interest in the second year)

If we observe the values of the interest in the 1st, 2nd and 3rd years, we observe that:

The interest in the nth year = $(100 + r)$% of (The interest in the $(n-1)^{th}$ year)

Using the above relation in our problem:

Interest in the 5th year = \$4,800

Interest in the 6th year = \$5,520

Thus, we have

$5,520 = (100 + r)\%$ of 4,800

$$=> 5,520 = \left(\frac{100 + r}{100}\right) \times 4,800$$

$$=> 5,520 = 48 \times (100 + r)$$

$$=> 115 = 100 + r$$

$$=> r = 15\%$$

The correct answer is Option D.

4.10 Functions

138. We have

$$f(p) = p^2 + \frac{1}{p^2}$$

$$\Rightarrow f\left(-\frac{1}{\sqrt{p}}\right) = \left(-\frac{1}{\sqrt{p}}\right)^2 + \frac{1}{\left(-\frac{1}{\sqrt{p}}\right)^2} = \frac{1}{p} + p$$

$$= \left(f\left(-\frac{1}{\sqrt{p}}\right)\right)^2 = \left(\frac{1}{p} + p\right)^2$$

$$= \frac{1}{p^2} + p^2 + 2 \times \frac{1}{p} \times p$$

$= p^2 + \frac{1}{p^2} + 2 = f(p) + 2$; by replacing the value of $p^2 + \frac{1}{p^2} = f(p)$

The correct answer is Option A.

139. $f(x) = -\frac{1}{x}$

$$\Rightarrow f(a) = -\frac{1}{a} = -\frac{1}{2}$$

$$\Rightarrow a = 2$$

Also, we have

$$f(ab) = -\frac{1}{ab} = \frac{1}{6}$$

$$\Rightarrow ab = -6$$

$$\Rightarrow b = -\frac{6}{a} = -\frac{6}{2}$$

$$\Rightarrow b = -3$$

The correct answer is Option D.

140. $f(x) = \sqrt{x} - 20$

$$\Rightarrow f(q) = \sqrt{q} - 20$$

Since $p = f(q)$, we have

$$p = \sqrt{q} - 20$$

=> $\sqrt{q} = p + 20$

Squaring both sides, we have

=> $q = (p + 20)^2$

The correct answer is Option A.

141. Let d, e and f be the hundreds, tens and units digits of K, respectively and g, h and i be the hundreds, tens and units digits of R, respectively.

Thus, we have

$f(K) = 2^d 3^e 5^f$

$f(R) = 2^g 3^h 5^i$

Since $f(K) = 18f(R)$, we have

$2^d 3^e 5^f = 18 \times 2^g 3^h 5^i$

=> $2^d 3^e 5^f = 2^{(g+1)} 3^{(h+2)} 5^i$

Since d, e, f, g, h, i are integers, comparing coefficients of 2, 3 and 5:

$d = g + 1$; $e = h + 2$ and $f = i$

Thus, we have

The three-digit number $R \equiv ghi$, where g is the digit in the hundreds place, h is the digit in the tens place and i is the digit in the tens place

=> $R = 100g + 10h + i$

The three-digit number $K \equiv def \equiv (g + 1)(h + 2)i$, where $(g + 1)$ is the digit in the hundreds place, $(h + 2)$ is the digit in the tens place and i is the digit in the tens place

=> $K = 100(g + 1) + 10(h + 2) + i$

=> $K = (100d + 10h + i) + 120$

=> $K = R + 120$

=> $K - R = 120$

The correct answer is Option E.

142. Working with the options one at a time:

Option A: $f(x) = 1 + x$
$=> f(1 - x) = 1 + (1 - x) = 2 - x \neq f(x)$ – Incorrect

Option B: $f(x) = 1 + x^2$
$=> f(1 - x) = 1 + (1 - x)^2 = 1 + (1 - 2x + x^2) = 2 - 2x + x^2 \neq f(x)$ – Incorrect

Option C: $f(x) = x^2 - (1 - x)^2$
$=> f(1 - x) = (1 - x)^2 - (1 - (1 - x))^2 = (1 - x)^2 - x^2 \neq f(x)$ – Incorrect

Option D: $f(x) = x^2(1 - x)^2$
$=> f(1 - x) = (1 - x)^2(1 - (1 - x))^2 = (1 - x)^2x^2 = f(x)$ – Correct

Since we already have the answer, we need not check Option E.

Verifying Option E, we would have had:

Option E: $f(x) = \frac{x^2}{1 - x}$

$=> f(1 - x) = \frac{(1 - x)^2}{1 - (1 - x)} = \frac{(1 - x)^2}{x} \neq f(x)$

The correct answer is Option D.

143. $f(x) = \frac{1}{x}$

$g(x) = \frac{x}{x^2 + 1}$

$=> f(g(x)) = \frac{1}{g(x)} = \frac{1}{\left(\frac{x}{x^2+1}\right)}$

$= \frac{x^2 + 1}{x}$

$= \frac{(x - 1)^2 + 2x}{x}$

$= \frac{(x - 1)^2}{x} + 2$

Since $x > 0$, the minimum value of the above expression will occur when the square term becomes zero (since a square term is always non-negative, the minimum possible value occurs when it is zero).

$=> (x - 1)^2 = 0 => x = 1$

Thus, the minimum value of $f(g(x)) = 0 + 2 = 2$

The correct answer is Option E.

144. $f(x) = \frac{10x}{1-x}$

$=> f(3) = \frac{10 \times 3}{1-3} = -15$

Thus, we have

$f(x) = \frac{1}{2} \times -15 = \frac{-15}{2}$

$=> \frac{10x}{1-x} = \frac{-15}{2}$

$=> 20x = -15 + 15x$

$=> 5x = -15$

$=> x = -3$

The correct answer is Option D.

145. We have

$3f(x) + 2f(-x) = 5x - 10 \ldots$(i)

Substituting $x = 1$ in equation (i):

$3f(1) + 2f(-1) = 5 - 10$

$=> 3f(1) + 2f(-1) = -5 \ldots$(ii)

Substituting $x = -1$ in equation (i):

$3f(-1) + 2f(1) = -5 - 10$

$=> 3f(-1) + 2f(1) = -15 \ldots$(iii)

Equation (ii) $\times$ 3 – Equation (iii) $\times$ 2:

$5f(1) = -15 - (-30) = 15$

$=> f(1) = 3$

The correct answer is Option D.

146. We know that

$D(t) = -10(t-7)^2 + 100$, where $0 \leq t \leq 12$

We need to find the value of t so that the value of $D(t)$ is maximum.

In the expression for $D(t)$, we have a negative term: $-10(t-7)^2$

We know that $(t-7)^2 \geq 0$ for all values of t (since it is a perfect square).

Thus, we have

$-10(t-7)^2 \leq 0$ for all values of t (multiplying with a negative reverses the inequality).

Thus, in order that $D(t)$ attains a maximum value, the term $-10(t-7)^2$ must be 0.

Thus, we have

$-10(t-7)^2 = 0$

$=> t = 7$

Thus, $D(t)$ attains a maximum value at $t = 7$ i.e. 7 hours past 12:00 am i.e. 7:00 am.

The correct answer is Option B.

4.11 Permutation & Combination & Probability

147. Since $C_3^5 = C_r^5$, $r = 3$, but there is no option as $r = 3$.

We know that $C_q^p = C_{p-q}^p$

$=> C_3^5 = C_r^5 = C_{5-r}^5$

$=> 3 = 5 - r$

$=> r = 2$

The correct answer is Option C.

148. This is a question on permutation with indistinguishable or identical objects.

We know that if there are n objects, out of which p objects are indistinguishable, then

Total number of way of arranging them = $\frac{n!}{p!}$

In this question, let's first assume that we use two BLACK and one RED dot, thus,

Total number of way of arranging them = $\frac{3!}{2!} = 3$.

Similarly, let's now assume that we use two RED and one BLACK dot, thus,

Total number of way of arranging them = $\frac{3!}{2!} = 3$.

There can be two more cases where we use all three BLACK or all three RED.
(Note: the question does not say that both colors must be used)

Total number of codes = 3 + 3 + 2 = 8.

The codes would be: RRB, RBY, RBB, BBR, BRB, BRR, BBB, & RRR.

The correct answer is Option D.

149. There are two different sizes and four different colors of mugs.

Packages having the same size and same color of mugs:

Number of ways in which a size can be chosen = $C_1^2 = \frac{2!}{(2-1)! \times 1!} = 2$ ways

Number of ways in which a color can be chosen = $C_1^4 = \frac{4!}{(4-1)! \times 1!} = 4$ ways

Thus, total number of such packages = $2 \times 4 = 8$.

Packages having the same size and different colors of mugs:

Number of ways in which a size can be chosen = $C_1^2 = \frac{2!}{(2-1)! \times 1!} = 2$ ways

Number of ways in which three different colors can be chosen = $C_3^4 = \frac{4!}{(4-3)! \times 3!} = 4$ ways

Thus, total number of such packages = $2 \times 4 = 8$

Thus, total number of different packages = $8 + 8 = 16$

The correct answer is Option C.

150. We know that there are six kinds of toppings and two kinds of breads for pizzas.

Since each pizza contains at least two kinds of toppings but not all kinds of toppings, the number of topping would be between 2 to 5.

Number of ways of selecting two kinds of toppings = $C_2^6 = \frac{6!}{4!2!} = 15$.

Number of ways of selecting three kinds of toppings = $C_3^6 = \frac{6!}{3!3!} = 20$.

Number of ways of selecting four kinds of toppings = $C_4^6 = C_2^6 = \frac{6!}{4!2!} = 15$.

Number of ways of selecting five kinds of toppings = $C_5^6 = C_1^6 = 6$.

Thus, number of possible sections of toppings = $15 + 20 + 15 + 6 = 56$.

Number of ways of selecting one kind of bread = $C_1^2 = 2$.

Thus, number of pizzas possible = $56 \times 2 = 112$.

The correct answer is Option E.

151. Number of one-letter codes:

The botanist can uniquely designate 26 plants (since there are a total of 26 letters).

Number of two-letter codes:

The first position can be assigned in 26 ways.

The second position can also be assigned in 26 ways (since the letters may be repeated).

Thus, total two-letter codes possible = $26 \times 26 = 676$.

Thus, using two-letter codes, the botanist can uniquely designate 676 plants.

Number of three-letter codes:

Each of the three positions can be assigned in 26 ways.

Thus, total three-letter codes possible = $26 \times 26 \times 26 = 17,576$.

Thus, using three-letter codes, the botanist can uniquely designate 17,576 plants.

Thus, total number of unique designations possible using one-, two- or three-letter codes = 26 + 676 + 17,576 = 18,278.

The correct answer is Option E.

Alternate approach:

There is a cheeky method for this question.

Number of one-letter codes: $26 \equiv$ units digit is 6.

Number of two-letter codes: $26 \times 26 \equiv$ units digit is 6.

Number of three-letter codes: $26 \times 26 \times 26 \equiv$ units digit is 6.

Thus, the units digit of the sum = $6 + 6 + 6 \equiv 8$.

Only Option E has the units digit as 8.

Note: This is not a holistic method and not to be used when two or more options are with same units digit.

152. We know that no subject is common in both groups.

Here events "Selection of one out of eight optional subjects from group one" and "Selection of two out of ten optional subjects from group two" are mutually exclusive or disjoint events; thus the total number of ways would be multiplied.

Number of ways of selecting one optional subject from eight subjects = $C_1^8 = 8$

Number of ways of selecting two optional subjects from ten subjects = $C_2^{10} = \dfrac{10 \times 9}{2 \times 1} = 45$.

Thus, total number of ways of selecting three subjects = $8 \times 45 = 360$.

The correct answer is Option D.

153. We need to form a four-digit code using the digits 1, 2, 3, 4, 5, 6, 7, 8, and 9 i.e. 9 possible digits.

Also, it is known that repetition of digits is not allowed.

The thousands position of the code can be filled using any of the nine digits in 9 ways.

The hundreds position of the code can be filled using any of the eight digits in 8 ways.

The tens position of the code can be filled using any of the remaining seven digits in 7 ways.

The units position of the code can be filled using any of the remaining six digits in 6 ways.

Thus, the number of distinct codes possible = $9 \times 8 \times 7 \times 6 = 3{,}024$.

The correct answer is Option C.

154. We need to form a four-digit code using the digits 0, 1, 2, 3, 4, 5, 6, 7, 8, and 9 i.e. 10 possible digits.

Also, it is known that repetition of digits is allowed, the code has to be an odd number, and the thousands' position of the code can't be 0.

The thousands' position of the code can be filled using any of the 9 digits (except 0) in 9 ways.

The hundreds position of the code can be filled using any of the 10 digits (since 0 can now be used and repetition is allowed) in 10 ways.

The tens position of the code can be filled using any of the 10 digits (since 0 can now be used and repetition is allowed) in 10 ways.

The units position of the code can be filled using any of the 5 odd digits 1, 3, 5, 7, or 9 in 5 ways.

Thus, the number of distinct codes possible = $9 \times 10 \times 10 \times 5 = 4{,}500$.

The correct answer is Option D.

155. Number of ways of selecting 4 sites out of 6

= (Number of ways in which any 4 of the 6 sites are selected without consideration to any constraint) – (Number of ways considering both A and B sites are selected)

Number of ways in which any 4 of the 6 sites may be selected without paying consideration to any restriction = C_4^6

$$= C_{(6-4)}^6 = C_2^6 = \frac{6 \times 5}{2 \times 1}$$

$= 15$

Number of ways in which 4 of the 6 sites may be selected so that both sites A and B are selected (i.e. two more sites to be selected from the remaining four) = $C_{(4-2)}^{(6-2)}$

$= C_2^4 = \frac{4 \times 3}{2 \times 1}$

= 6; the above 6 ways do not satisfy the given restriction

Thus, the number of ways in which 4 of the 6 sites can be selected so that both A and B sites are not selected simultaneously = $15 - 6 = 9$

The correct answer is Option D.

156. Imran has 4 Math, 5 Physics, and 6 Chemistry books.

To select: 4 books such that the selection has at least one book of each subject.

The selections can be done in following ways.

(1) 2 Math, 1 Physics and 1 Chemistry books:
# of ways = $C_2^4 \times 5 \times 6 = \left(\frac{4 \times 3}{1 \times 2}\right) \times 5 \times 6 = 180$

(2) 1 Math, 2 Physics and 1 Chemistry books:
# of ways = $4 \times C_2^5 \times 6 = 4 \times \left(\frac{5 \times 4}{1 \times 2}\right) \times 6 = 240$

(3) 1 Math, 1 Physics and 2 Chemistry books:
# of ways = $4 \times 5 \times C_2^6 = 4 \times 5 \times \left(\frac{6 \times 5}{1 \times 2}\right) = 300$

Total number of possible selections = $180 + 240 + 300 = 720$

The correct answer is Option B.

157. Let the number of letters to be used be n.

The number of plants that can be identified using a single letter = $C_1^n = n$.

The number of plants that can be identified using two distinct letters = $C_2^n = \frac{n(n-1)}{2}$ (since the letters are to kept in alphabetic order, we must not order them or apply P_2^n.)

Thus, total number of plants that can be identified if we attempt to have all the 15 codes that are either one-letter code or two-letter codes

$= \left(n + \frac{n(n-1)}{2}\right)$

Since we need to have at least 15 identifications, we have

$$n + \frac{n(n-1)}{2} \geq 15$$

$$=> \frac{2n + n(n-1)}{2} \geq 15$$

$$=> \frac{n(2 + (n-1))}{2} \geq 15$$

$$=> \frac{n(n+1)}{2} \geq 15$$

$$=> n(n+1) \geq 30$$

Working with the options and starting with the least value of n, we see that $n = 5$ satisfies the above inequality.

The correct answer is Option C.

158. We have to select 1 student each from classes A, B, and C each having 30 students and 2 students from Class D having 20 students.

Number of ways of selecting 1 student from Class A = $C_1^{30} = 30$.

Number of ways of selecting 1 student from Class B = $C_1^{30} = 30$.

Number of ways of selecting 1 student from Class C = $C_1^{30} = 30$.

Number of ways of selecting 2 student from Class D = $C_2^{20} = \frac{20 \times 19}{2 \times 1} = 190$.

As all four events are independent of each other, the number of ways of forming the team = $30 \times 30 \times 30 \times 190$

= 5,130,000

The correct answer is Option E.

159. Total number of stocks available: 4 Information Technology, 5 Retail, and 3 e-commerce stocks

Number of stocks to be selected: 2 Information Technology, 4 Retail, and 2 e-commerce stocks

Number of ways of selecting 2 out of 4 Information Technology stocks = $C_2^4 = \frac{4 \times 3}{2 \times 1} = 6$

Number of ways of selecting 4 out of 5 Retail stocks = $C_4^5 = C_{(5-4)}^5 = C_1^5 = 5$

Number of ways of selecting 2 out of 3 e-commerce stocks = $C_2^3 = C_{(3-2)}^3 = C_1^3 = 3$

Thus, the total number of ways of selecting the required number of stocks

$= 6 \times 5 \times 3$ (we multiply since the above selections are independent of one another)

$= 90$

The correct answer is Option D.

160. Total number of people present $= 3 \times 8 = 24$.

Number of handshakes if the delegates shook hands with every person other than those from his or her own company

= (Number of handshakes without any constraint) – (Number of handshakes with people from their own company)

Number of handshakes without any constraint $= C_2^{24} = \frac{24 \times 23}{2 \times 1} = 276$

(Since a handshake requires two people, we need to **select** any 2 people from the 24 for a handshake)

Number of handshakes with people from their own company $= 8 \times C_2^3 = 8 \times \frac{3 \times 2}{2 \times 1} = 24$

(We need to select any 2 delegates from 3 delegates of the same company, for each of the 8 companies)

Thus, the number of handshakes if the delegates shook hands with every person other than those from his or her own company

$= 276 - 24$

$= 252$

The correct answer is Option C.

Alternate approach:

Let us calculate how many handshake a particular person does.

There are a total of $3 \times 8 = 24$ delegates.

A particular person would have to handshake $24 - 3 = 21$ times. This excludes the person itself and his company's two collegaues.

Thus, all the delegates would do $\frac{21 \times 24}{2} = 252$ handshakes.

We divided the total number of handshakes by '2' because one handshake involving two delegates should be counted as one handshake and not two.

161. We need to select three digits for the code.

The digits to be used are from 0 to 9, thus there are 10 possible digits.

Since the first digit cannot be 0 or 9, number of possibilities for the first digit $= (10 - 2) = 8$.

Since the second digit can only be 0 or 9, number of possibilities for the second digit = 2.

Let us ignore the restriction for the third digit.

Thus, number of possibilities for the third digit = 10.

Thus, total number of codes possible (ignoring the condition for the third digit) $= 8 \times 2 \times 10 = 160$

In the above codes, there are a few codes that are unacceptable since they violate the condition for the third digit.

The codes which violate the condition for the third digit are of the form (a99), where a is the first digit and both second and third digits are simultaneously 9.

The number of such codes equals the number of possibilities for the first digit, i.e. 8.

Thus, the number of codes possible without violating any of the given conditions $= 160 - 8 = 152$

The correct answer is Option A.

162. There are a total of 30 marbles, out of which 15 are yellow.

Probability that both marbles are yellow $= \frac{\text{Number of ways of drawing two yellow marbles}}{\text{Number of ways of drawing any two marbles}}$

$$=> \frac{C_2^{15}}{C_2^{30}} = \frac{\frac{15\times14}{1\times2}}{\frac{30\times29}{1\times2}} = \frac{15 \times 14}{30 \times 29} = \frac{7}{29}$$

The correct answer is Option B.

163. In a basket, out of 12 balls, seven are red and five are green.

Number of ways we can select three balls from 12 balls $= C_3^{12} = \frac{12 \times 11 \times 10}{3 \times 2 \times 1} = 220$

Number of ways of selecting two red balls and one green ball

$$= C_2^7 \times C_1^5 = \frac{7 \times 6}{1 \times 2} \times 5 = 21 \times 5 = 105$$

So, the required probability = $\frac{C_2^7 \times C_1^5}{C_3^{12}} = \frac{105}{220} = \frac{21}{44}$

The correct answer is Option D.

164. The number of 6–member committees that can be formed from the 21 members = C_6^{21}

$= \frac{21 \times 20 \times 19 \times 18 \times 17 \times 16}{6 \times 5 \times 4 \times 3 \times 2 \times 1}$

The number of 5–member committees that can be formed from the 21 members = C_5^{21}

$= \frac{21 \times 20 \times 19 \times 18 \times 17}{5 \times 4 \times 3 \times 2 \times 1}$

Thus, the required ratio

$$= \frac{\left(\frac{21 \times 20 \times 19 \times 18 \times 17 \times 16}{6 \times 5 \times 4 \times 3 \times 2 \times 1}\right)}{\left(\frac{21 \times 20 \times 19 \times 18 \times 17}{5 \times 4 \times 3 \times 2 \times 1}\right)}$$

$= \frac{16}{6} = \frac{8}{3} = 8$ to 3

The correct answer is Option C.

165. There are two offices to which the four employees need to be assigned.

Thus, the number of offices for each employee to choose = 2 (since each employee can be assigned to any of the two offices)

Thus, the total number of ways of assigning the four employees = $2 \times 2 \times 2 \times 2 = 2^4 = 16$.

Note: For n objects, each with r options, the total number of options = r^n

The correct answer is Option E.

166. We need to select three male officers from five male officers.

The number of ways of achieving it = $C_3^5 = \left(\frac{5 \times 4 \times 3}{3 \times 2 \times 1}\right) = 10$.

We also need to select two female officer from three female officers.

The number of ways of achieving it = $C_2^3 = C_1^3 = 3$.

Thus, the number of ways in which three male officers and two female officer can be selected = $10 \times 3 = 30$.

The correct answer is Option D.

167. Given that,

Probability that Stock X increases in value = P(A) = 0.4;

Probability that Stock Y increases in value = P(B) = 0.6

Exactly one of the stocks would increase in value if:

Stock A increases AND Stock Y does not

OR

Stock B increases AND Stock X does not

Probability that Stock X does not increase in value = $P(\overline{A}) = 1 - 0.4 = 0.6$
Probability that Stock Y does not increase in value = $P(\overline{B}) = 1 - 0.6 = 0.4$

Thus, the required probability = $P(A) \times P\left(\overline{B}\right) + P(B) \times P\left(\overline{A}\right)$

$= 0.4 \times 0.4 + 0.6 \times 0.6$

$= 0.16 + 0.36 = 0.52$

The correct answer is Option C.

168. Since probability of landing on heads or tails is the same, each must be $\frac{1}{2}$ => $P(H) = P(T) = \frac{1}{2}$

Probability that the coin will land on heads at least once on two tosses

= 1− Probability that it will not land on heads at all

= 1− Probability that it will land on tails on both occasions

= 1− (Probability that the first toss will show tails AND the second toss will show tails)

$= 1 - P(T) \times P(T)$

$= 1 - \frac{1}{2} \times \frac{1}{2}$

$= 1 - \frac{1}{4}$

$= \frac{3}{4}$

The correct answer is Option E.

169. Each question has two options of which only one is correct.

Thus, the probability of randomly guessing an answer and getting it correct = $\frac{1}{2}$.

Thus, the probability of randomly guessing answers to all X questions and getting them correct

$= \frac{1}{2} \times \frac{1}{2} \times \frac{1}{2} \times \ldots X$ times

$= \left(\frac{1}{2}\right)^X$

Thus, we have

$\left(\frac{1}{2}\right)^X < \frac{1}{500}$

$=> 2^X > 500$

We observe from the options that $2^9 = 512$, which just exceeds 500.

Thus, the least value of $n = 9$.

The correct answer is Option B.

170. We know that the first three balls were black.

Thus, there are $(20 - 3) = 17$ balls left.

Of these, number of black balls $= 10 - 3 = 7$

We need the next two balls to be black.

Thus, we need to select 2 black balls from the remaining 7 black balls.

Number of ways in which the above case can be achieved (favorable cases) $= C_2^7$

$= \frac{7 \times 6}{2 \times 1} = 21$

Number of ways in which two balls can be selected from the remaining 17 balls (total cases) $= C_2^{17} = \frac{17 \times 16}{2 \times 1} = 136$

Thus, required probability $= \frac{\text{Favorable cases}}{\text{Total cases}}$

$= \frac{21}{136}$

The correct answer is Option A.

171. We need to select two balls one at a time.

On drawing 2 balls, one white and one blue can be obtained if:

The first ball is white AND the second ball is blue

OR

The first ball is blue AND the second ball is white

Since the balls are not replaced after drawing, after the first draw, the total number of balls would be 1 less than what was present initially, i.e. $(16 - 1) = 15$.

Thus, the required probability =

p(The first ball is white AND the second ball is blue)

OR

p(The first ball is blue AND the second ball is white)

$= \{p(1^{st}\text{ ball white}) \times p(2^{nd}\text{ ball blue})\} + \{p(1^{st}\text{ ball blue}) \times p(2^{nd}\text{ ball white})\}$

$$= \left\{\frac{4}{16} \times \frac{3}{15}\right\} + \left\{\frac{3}{16} \times \frac{4}{15}\right\}$$

$$= \frac{1}{20} + \frac{1}{20}$$

$$= \frac{1}{10}$$

The correct answer is Option C.

172. Probability that at least one refrigerator is single-door

= 1 – Probability that none is single-door refrigerator

Probability of selecting two refrigerators such that none is single-door refrigerator

= Probability of selecting two double-door refrigerators

$$= \left(\frac{\text{Number of ways of selecting 2 doubledoor from 6 sets}}{\text{Number of ways of selecting 2 refrigerators from 8 sets}}\right)$$

$$= \frac{C_2^6}{C_2^8}$$

$$= \frac{\left(\frac{6 \times 5}{2 \times 1}\right)}{\left(\frac{8 \times 7}{2 \times 1}\right)}$$

$= \frac{15}{28}$

Thus, the required probability $= 1 - \frac{15}{28} = \frac{13}{28}$.

The correct answer is Option D.

173. Let the number of red balls = r

Number of white balls = 9

Thus, total number of balls = $(r + 9)$

Thus, probability that both balls would be white

$$= \left(\frac{\text{Number of ways of selecting 2 white balls form 9 white balls}}{\text{Number of ways of selecting 2 balls from } (r+9) \text{ balls}}\right)$$

$$= \frac{C_2^9}{C_2^{(r+9)}}$$

$$= \frac{\frac{9 \times 8}{2 \times 1}}{\frac{(r+9) \times (r+9-1)}{2 \times 1}}$$

$$= \frac{72}{(r+9) \times (r+8)}$$

Thus, we have

$$\frac{72}{(r+9) \times (r+8)} = \frac{6}{11}$$

$=> (r + 9) \times (r + 8) = 12 \times 11$

We see that the LHS and the RHS are product of two consecutive numbers, thus $(r + 9) = 12 => r = 3$

Thus, the total number of balls = 3 + 9 = 12

The correct answer is Option C.

174. The probability that a card falls = $P(F) = 0.05$

Thus, the probability that a card does not fall = $P\left(\overline{F}\right) = 1 - 0.05 = 0.95$

Thus, the probability that none of the 12 card falls = 0.95^{12}

For the pyramid to collapse, at least one card needs to fall.

Thus, the probability that at least one card falls = $1 - 0.95^{12}$

The correct answer is Option C.

175. Total number of candies = 12

Number of orange flavored candies = 4

Thus, the number of non-orange flavored candies = 12 - 4 = 8

Thus, the kid has to pick 2 candies out of the 8 non-orange flavored candies.

Thus, number of ways (favorable cases)

$$= C_2^8 = \frac{8 \times 7}{2 \times 1} = 28$$

Number of ways in which the kid can pick 2 candies from the 12 candies (total cases)

$$= C_2^{12} = \frac{12 \times 11}{2 \times 1} = 66$$

Thus, the required probability

$$= \frac{\text{Favorable cases}}{\text{Total no. of cases}} = \frac{28}{66} = \frac{14}{33}$$

The correct answer is Option C.

176. If Suzy returns at the end of the day 3, she would have stayed for 3 days.

Since it is given that she would return home at the end of the first day it rained, it must not have rained on day 1 and day 2 but rained on day 3.

Probability of rain on each day = $P\,(R) = 0.25$

=> Probability that there is no rain on a particular day

$$= P\left(\overline{R}\right) = 1 - P\,(R)$$

$$= 1 - 0.25$$

$$= 0.75$$

Thus, probability that she return at the end of day on day 3

= Probability that there is no rain on day 1 AND no rain on day 2 AND rain on day 3

$$= P\left(\overline{R}\right) \times P\left(\overline{R}\right) \times P\,(R)$$

$$= 0.75 \times 0.75 \times 0.25 = \frac{9}{64}$$

The correct answer is Option B.

4.12 Sets

177. Number of students on the committee G1 = 10.

As no member of G1 is in either of the other two groups, the above 10 students belong to only G1.

However, there may be an overlap with the students of G2 and G3.

Number of students in G2 = 10.

Number of students in G3 = 6.

We would get the greatest number of students who would not be in any of the groups if there is maximum overlap between the students of G2 and G3.

The maximum overlap between the students of G2 and G3 would be the minimum of the number of students in the two groups i.e. minimum of 6 and 10 = 6.

Thus, we have

$G2 \cap G3 = 6$

Thus, the number of students in G2 or G3 = $G2 + G3 - G2 \cap G3 = 10 + 6 - 6 = 10$

Thus, total number of students belonging to one of more groups

= G1 + G2 or G3 = 10 + 10 = 20.

Thus, maximum number of students who don't belong to any group = 30 − 20 = 10.

The above information can be represented in a Venn-diagram as shown below:

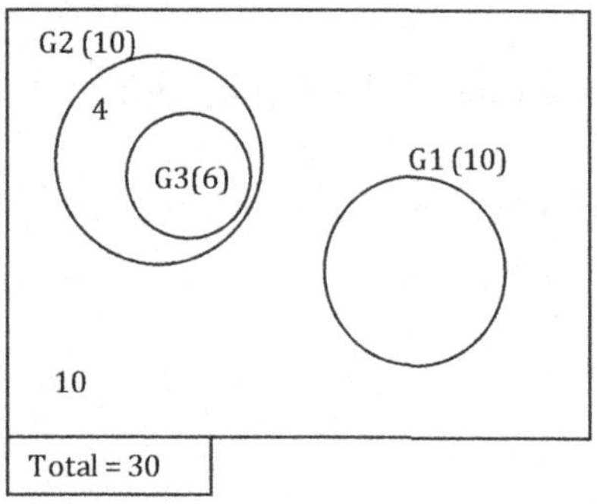

The correct answer is Option D.

178. Total number of dresses = 1,000

Number of traditional dresses = 250

Number of contemporary dresses = 750

Number of dresses for males = Number of dresses for females = $\frac{100}{2}$ = 500

Number of traditional dresses for males = 100

Let us represent the above information using a Venn-diagram, as shown below:

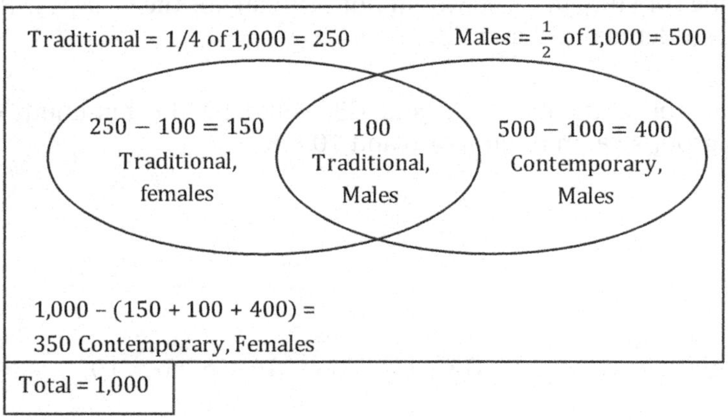

Thus, from the above Venn-diagram, we have

Number of contemporary dresses for females = 350

The correct answer is Option D.

179. We know that 95 percent students used a computer to play games or write reports.

Thus, (100 – 95) = 5% students did not use a computer for either of these purposes.

Let us represent the above information using a Venn-diagram, as shown below:

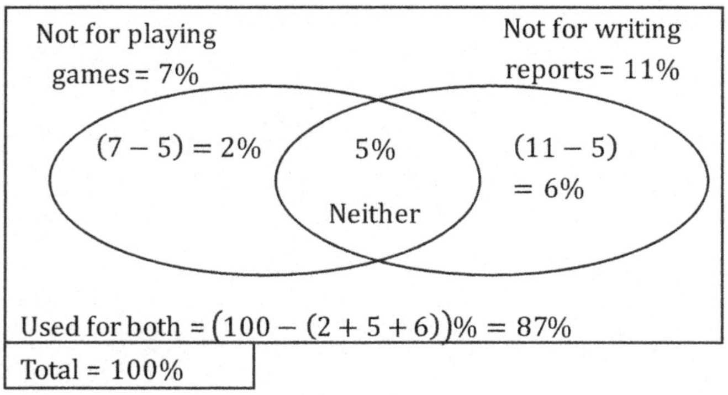

Thus, the percent of students who did use a computer both to play games and to write reports = 87%.

The correct answer is Option D.

180. Let us represent the given information using a Venn-diagram, as shown below:

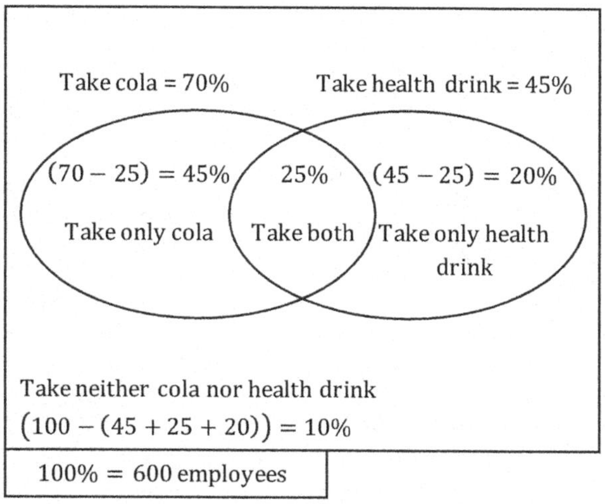

Thus, the number of employees who take neither cola nor health drink

= 10% of 600

= 60

The correct answer is Option B.

181. Let us represent the given information using a Venn-diagram, as shown below:

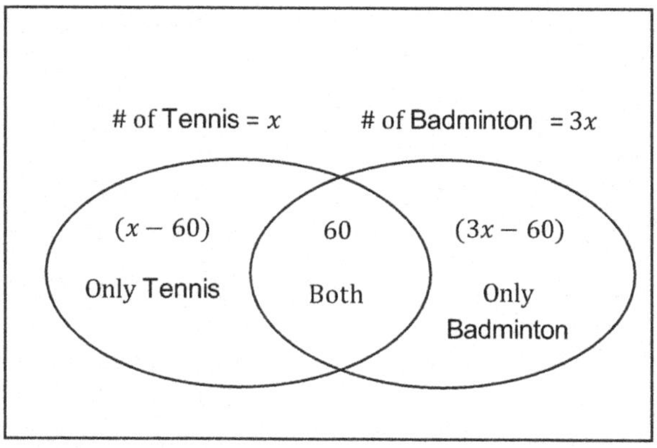

We know that the number of students who play both sports is thrice the number of students who play only Tennis.

Thus, we have

$60 = 3(x - 60)$

$=> x = 80$

Thus, the number of students play only Badminton $= 3x - 60$

$= 180$

The correct answer is Option D.

4.13 Statistics & Data Interpretation

182. We know that the median is the middle-most value of any series/data set, but we do not know the value of x, so we cannot calculate exact value of Median; however we can surely find its range.

- Case 1:
  If x is smallest, the series would be x, 15, 20, 25 and median = average of 15 & 20 = 17.5–smallest median value.
- Case 2:
  If x is largest, the series would be 15, 20, 25, x and median = average of 20 & 25 = 22.5–largest median value.

Thus, the median would lie between 17.5 & 22.5, inclusive. Since the values in only Statements I & II are in the range, Option C is correct.

The correct answer is Option C.

183. The question asks the number of scores > (Mean + SD)?

Mean + 1× SD = 73 + 1 × 20.50 = 93.50.

It is clear that three scores (95, 96, and 97) are greater than 93.50.

The correct answer is Option D.

184. Let the sum of the 19 numbers other than n be s.

Thus, we have

$$n = 4 \times \left(\frac{s}{19}\right)$$

$$=> s = \frac{19n}{4}$$

Thus, the sum of all the 20 numbers in the list $= s + n = \frac{19n}{4} + n = \frac{23n}{4}$

Thus, the required fraction $= \frac{n}{\frac{23n}{4}} = \frac{4}{23}$

The correct answer is Option B.

185. We have

$$w \leq \frac{3 + 8 + w}{3} \leq 3w$$

=> $3w \leq 11 + w \leq 9w$

=> $2w \leq 11 \leq 8w$

=> $2w \leq 11$ and $8w \geq 11$

=> $w \leq \frac{11}{2} = 5\frac{1}{2}$

and

=> $w \geq \frac{11}{8} = 1\frac{3}{8}$

Since w is an integer, possible values of w are: 2, 3, 4 or 5,

Thus, there are four possible values of w.

The correct answer is Option B.

186. Let the seven distinct positive integers be p, q, v, w, x, y and z such that $p > q > v > w > x > y > z$.

Thus, we need to find the least possible value of the largest among the seven, i.e. p.

Let's assume that w, the middle-most number = average of the seven numbers = 14.

Now to maintain the average equals to 14, increase v, q, & p by 1, 2, & 3, respectively and decrease x, y, & z by 1, 2, & 3, respectively.

Thus, $p = 14 + 3 = 17$ – the least possible value of the greatest of the seven numbers

The correct answer is Option B.

187. We have

$$\frac{x + y + 10}{3} = \frac{x + y + 10 + 20}{4}$$

=> $4x + 4y + 40 = 3x + 3y + 90$

=> $x + y = 50$

The correct answer is Option B.

188. If the 13 different integers are arranged in order, the median is the $\left(\frac{13+1}{2}\right)^{th} = 7^{th}$ integer.

Thus, the 7th integer is 20.

Since we need to find the maximum possible integer (given a range of 20), we need to have the maximum value of the least integer as well.

Since the integers are distinct, we can have the first 7 integers as:

$(20-6) = 14$, $(20-5) = 15$, $(20-4) = 16$, $(20-3) = 17$, $(20-2) = 18$, $(20-1) = 19$, and 20

Thus, the maximum possible value of the smallest integer = 14.

Since the range is 20, the value of the greatest integer = 14 + 20 = 34.

The correct answer is Option D.

189. We have Mean = $\left(\frac{\text{Sum of the terms}}{\text{Total number of terms}}\right)$

$=> \frac{1+2+3+4+5+6+x}{7} = \frac{2x}{7}$

$=> \frac{21+x}{7} = \frac{2x}{7}$

$=> 3 + \frac{x}{7} = \frac{2x}{7}$

$=> 3 = \frac{2x}{7} - \frac{x}{7}$

$=> 3 = \frac{x}{7}$

$=> x = 21$

The correct answer is Option E.

190. We know that there are x employees.

Since the median salary is the 22nd salary, no two salaries are the same, and the 22nd salary is there in the list, there would be 21 salaries that are less than the 22nd salary and 21 salaries that are greater than the 22nd salary.

This implies that there are 21 + 1 + 21 = 43 salaries in the list or there are a total of 43 employees.

Thus, the average salary of the 43 employees

$= \frac{\text{Total salary}}{\text{Number of employees}}$

$= \$\left(\frac{860,000}{43}\right)$

$= \$20,000$

The correct answer is Option B.

191. We can calculate the total number of watts of electricity consumed as shown below:

Appliance	Number of hours in use	Number of watts of electricity used per hour	Total number of watts of electricity used
Computer	4	105	$4 \times 105 = 420$
Music system	2	90	$2 \times 90 = 180$
Refrigerator	2	235	$2 \times 235 = 470$
LED TV	2	150	$2 \times 150 = 300$
Total	$4 + 2 + 2 + 2 = 10$ hours		$420 + 180 + 470 + 300 = 1{,}370$ watts

Thus, the average number of watts of electricity used per hour per appliance

$= \frac{1,370}{10} = 137$ watts per hour.

The correct answer is Option C.

192. We know that the distribution is symmetric about the mean.

Thus, the percent of the distribution equidistant from the mean on either side of it is the same.

Let the percent of the distribution less than $(\bar{x} + d)$ be $x\%$.

Thus, the percent of the distribution more than $(\bar{x} - d)$ is also $x\%$.

The situation is shown in the diagram below.

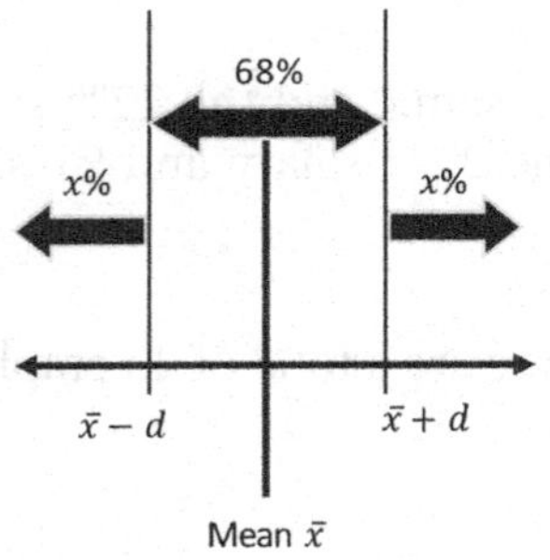

Thus, we have

$x\% + 68\% + x\% = 100\%$

$=> x = 16\%$

Thus, the percent of the distribution greater than $(\bar{x} - d) = x + 68 = 16 + 68 = 84\%$.

The correct answer is Option D.

4.14 Linear Equations

193. Let us recall the property of two-digit number:

"Difference between a particular two digit number and the number obtained by interchanging the digits of the same two digit number is always 9 times the difference between the digits."

Thus, the difference between actual amount and reversed amount = 63

= 9 × difference between the digits

=> Difference between the digits = $\frac{63}{9} = 7$

The difference between the digits of the number 7 is satisfied only by Option E.

The correct answer is Option E.

Alternate approach 1:

If we consider correct amount as $[xy] = 10x + y$, then interchanged amount becomes $[yx] = 10y + x$.

According to the given condition, difference between the new amount and the original amount is 63 cents.

=> $(10x + y) - (10y + x) = 63$

=> $10x - x - 10y + y = 63$

=> $9x - 9y = 63$

On dividing by 9, we have

$x - y = 7$.

Thus, the difference between the digits is 7, which is satisfied only by Option E.

Alternate approach 2:

Since the cash register contained 63 cents more than it should have as a result of this error, this implies that the tens digit of the correct amount must be greater than its units digit.

Only two options can qualify. Let us analyze them:

(C) 73: $73 - 37 = 36 \neq 63$

(E) 92: $92 - 29 = 63$; correct answer

194. Let the number of screwdrivers and spanners purchased be a and b, respectively.

Thus, total cost of the items = $\$(11a + 3b)$

Thus, we have

$11a + 3b = 109$

$$b = \frac{109 - 11a}{3} = \frac{108 + 1 - 11a}{3} = \frac{108}{3} + \frac{1 - 11a}{3} = 36 + \frac{1 - 11a}{3}$$

It is clear that a and b are positive integers. Thus, $(1 - 11a)$ must be a multiple of 3.

There can be a few possible values of a, two such values are $a = 2$, 5 and 8.

For $a = 3$, we have $b = 36 + \frac{1 - 11 \times 2}{3} = 36 - 7 = 29$. Thus, $a + b = 2 + 29 = 31$.

Since 31 is not one of the options, we must try with $a = 5$.

For $a = 5$, we have $b = 36 + \frac{1 - 11 \times 5}{3} = 36 - 18 = 18$. Thus, $a + b = 5 + 18 = 23$.

Since 23 is not one of the options, we must try with $a = 8$.

For $a = 8$, we have $b = 36 + \frac{1 - 11 \times 8}{3} = 36 - 29 = 7$. Thus, $a + b = 8 + 7 = 15$.

Thus, the total number of items purchased could be $8 + 7 = 15$.

The correct answer is Option C.

195. We have

$x + y + z = 2 \ldots$ (i)
$x + 2y + 3z = 6 \ldots$ (ii)

Since z has no role to play, let's eliminate it.

Multiplying (i) by 3 and subtracting (ii) from the result:

$3 \times (x + y + z) - (x + 2y + 3z) = 3 \times 2 - 6$

$\Rightarrow 2x + y = 0$

$\Rightarrow 2x = -y$

$\Rightarrow \frac{x}{y} = -\frac{1}{2}$

The correct answer is Option A.

196. Let the number of notebooks sold last month be n.

Since the number of books sold was 2 more than the notebooks, the number of books sold = $(n+2)$.

Selling price of each book = \$25.

Selling price of each notebook = \$15.

Thus, total sales revenue = $\$(15n + 25(n+2))$.

Thus, we have

$15n + 25(n+2) = 490$

$40n = 490 - 50$

$=> n = \frac{440}{40} = 11$

Thus, the number of notebooks sold = 11.

Thus, the number of books sold = 11 + 2 = 13.

Thus, the total number of books and notebooks sold = 11 + 13 = 24.

The correct answer is Option C.

Alternate approach:

We know that the number of books sold is 2 more than the number of notebooks sold.

The price of 2 books = $\$(25 \times 2) = \$50$.

Removing this from the total, i.e. \$490, we are left with $\$(490 - 50) = \$440$.

This amount was obtained by selling equal numbers of books and notebooks.

Total price of one book and one notebook = $\$(25 + 15) = \$40$.

Thus, number of items sold for \$440

$= \frac{440}{40} \times 2 = 22$

Thus, total number of books and notebooks sold = 2 + 22 = 24.

4.15 Quadratic Equations & Polynomials

197. We have

$a = \sqrt{8ab - 16b^2}$

Squaring both sides:

$a^2 = 8ab - 16b^2$

$=> a^2 - 8ab + 16b^2 = 0$

$=> (a - 4b)^2 = 0$

$=> a - 4b = 0$

$=> a = 4b$

The correct answer is Option D.

198. Given that,

$(x - 2)^2 = 9$

$=> x - 2 = \pm 3$

$=> x = 2 \pm 3$

$=> x = 5 \text{ OR } -1$

Given that,

$(y - 3)^2 = 25$

$=> y - 3 = \pm 5$

$=> y = 3 \pm 5$

$=> y = 8 \text{ OR } -2$

The minimum value of $\left(\frac{x}{y}\right)$ will be that value with the greatest magnitude of x, least magnitude of y and exactly one between of x and y being negative in sign.

Thus, we have

$x = 5, y = -2 => \frac{x}{y} = -\frac{5}{2}$

The maximum value of $\left(\frac{x}{y}\right)$ will be that value with the greatest magnitude of x, least magnitude of y and both x and y being simultaneously positive or negative in sign.

Thus, we have

$x = 5, y = 8 \Rightarrow \frac{x}{y} = \frac{5}{8}$

OR

$x = -1, y = -2 \Rightarrow \frac{x}{y} = \frac{1}{2}$

Between $\frac{5}{8}$ and $\frac{1}{2}$, the fraction $\frac{5}{8}$ is greater.

Thus, the required difference

$= \frac{5}{8} - \left(-\frac{5}{2}\right)$

$= \frac{5}{8} + \frac{5}{2}$

$= \frac{25}{8}$

The correct answer is Option E.

199. Given that,

$2x + 3y + xy = 12$

$\Rightarrow 2x + 3y + xy + 6 = 12 + 6 = 18$; (adding the product of the coefficients of x and y to both sides)

$\Rightarrow (2x + 6) + (xy + 3y) = 18$

$\Rightarrow 2(x + 3) + y(x + 3) = 18$

$\Rightarrow (x + 3)(y + 2) = 18$

Since x and y are positive integers, we must have:

$x + 3 > 3$, and

$y + 2 > 2$

Possible ways of getting 18 are: $1 \times 18, 2 \times 9$, and 3×6

Thus, the only possible solution is:

$x + 3 = 6 => x = 3$, and

$y + 2 = 3 => y = 1$

Thus, we have $x + y = 3 + 1 = 4$.

The correct answer is Option B.

Alternate approach:

Given that both x and y are positive integers, we can solve this question by hit and trial approach, too. Let's plug-in few probable positive integer values for x in the equation $2x + 3y + xy = 12$ and see which value renders a positive integer value for y.

We see that for $x = 1$ & 2, we get non-integer values for y; and for $x = 3$, we get $y = 1$, a positive integer value. Thus, $x + y = 3 + 1 = 4$

200. We know that

$h = -3(t - 10)^2 + 250 \ldots$ (i)

We need to first find the value of t such that the value of h is maximum.

In the expression for h, we have a term $-3(t - 10)^2$

We know that $(t - 10)^2 \geq 0$ for all values of t (since it is a perfect square).

Thus, we have

$-3(t - 10)^2 \leq 0$ for all values of t (multiplying with a negative reverses the inequality).

Thus, in order that h attains a maximum value, the term $-3(t - 10)^2$ must be 0.

Thus, we have $-3(t - 10)^2 = 0$

$=> t = 10$

Thus, h attains a maximum value at $t = 10$ seconds

Thus, 7 seconds after the maximum height is attained, i.e. at $t = 10 + 7 = 17$, we have the corresponding value of h (in feet) as:

$h = -3(t - 10)^2 + 250$

$= -3 \times 49 + 250$; substituting the value of $t = 17$ and solving

$= 103$ feet

The correct answer is Option B.

4.16 Inequalities

201. The possible extreme scenarios are shown in the diagrams below:

(1) Maximum distance away from hostel:

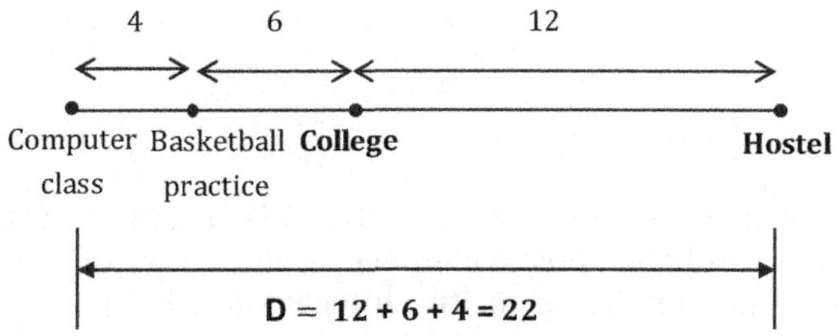

(2) Minimum distance away from hostel:

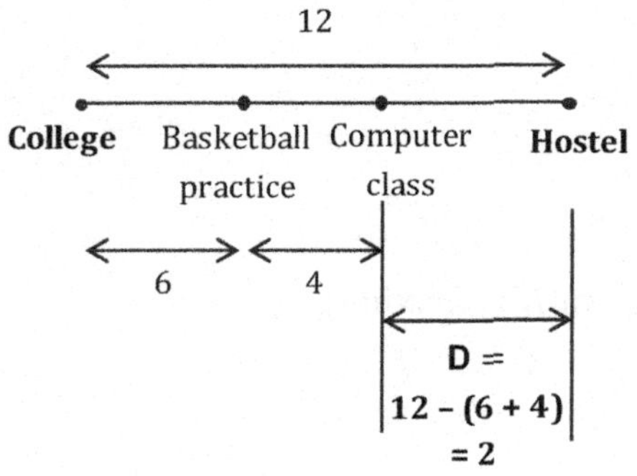

Thus, the maximum value of D is 22 and minimum value of D is 2

$=> 2 \leq D \leq 22$

The correct answer is Option D.

202. As per given inequality: $|b| \leq 12$, value of 'b' ranges from '-12' to '$+12$'. So, by putting these values in first equation: $2a + b = 12$, we can form a table of consistent values of a & b.

a	b	a	b
0	12	Cont...	
1	10	7	−2
2	8	8	−4
3	6	9	−6
4	4	10	−8
5	2	11	−10
6	0	12	−12

So a total of 13 ordered pairs are possible.

The correct answer is Option E.

Alternate approach:

We see that the value of b ranges from -12 to $+12$; this follows that b can have 25 number of integer values.

Now let us see how many integer values a can have.

$2a + b = 12 => a = \frac{12 - b}{2}$;

$=> a = 6 - \frac{b}{2}$

We see that for a to be an integer, $\frac{b}{2}$ must be an integer; this follows that b must be an even number.

Out of 25 possible values of b, 13 are even (including 0); so for a to be an integer, the set of arrangement can only have 13 ordered pairs.

203. Let the cost of an pencil be $\$x$ and the cost of a pen be $\$y$.

Thus, we need to find the range within which $(5x + 7y)$ lies.

We have

$3.60 < 15x < 4.80$

$=> \frac{3.60}{3} < \frac{15x}{3} < \frac{4.80}{3}$

$=> 1.20 < 5x < 1.60 \ldots$ (i)

Also, we have

$33.30 < 21y < 42.90$

$=> \frac{33.30}{3} < \frac{21y}{3} < \frac{42.90}{3}$

$=> 11.10 < 7y < 14.30 \ldots$ (ii)

Adding (i) and (ii):

$12.30 < 5x + 7y < 15.90$

Thus, the correct answer is Option D.

204. Working with the options one at a time:

Comparing options A and B:

$(xy)^2 = x^2y^2 < x^2$ (since $0 < y < 1 => y^2$ is a fraction between 0 and 1)

Thus, Option B cannot have the greatest value.

Comparing options A and C:

$$\left(\frac{x}{y}\right)^2 = \frac{x^2}{y^2} = x^2 \times \left(\frac{1}{y^2}\right) > x^2 \text{ (since } 0 < y < 1 => 0 < y^2 < 1 => \frac{1}{y^2} > 1\text{)}$$

Thus, Option A cannot have the greatest value.

Comparing options C and D:

$$\frac{x^2}{y} = \frac{x^2}{y^2} \times y = \left[\left(\frac{x}{y}\right)^2 \times y\right] < \left(\frac{x}{y}\right)^2 \text{ (since } 0 < y < 1 => y \text{ is a fraction between 0 and 1)}$$

Thus, Option D cannot have the greatest value.

Comparing options C and E:

$$x^2y = \frac{x^2}{y^2} \times y^3 = \left(\frac{x}{y}\right)^2 \times y^3 < \left(\frac{x}{y}\right)^2 \text{ (since } 0 < y < 1 => y^3 \text{ is a fraction between 0 and 1)}$$

Thus, Option E cannot have the greatest value.

The correct answer is Option C.

Alternate approach:

Since $x < 0$ and $0 < y < 1$ must be true for all the values of x & y; let us assume convenient, smart values of x & y.

Say $x = -1$ & $y = \frac{1}{2}$

Let us calculate the values of each option.

(A) $x^2 = (-1)^2 = 1$

(B) $(xy)^2 = \left(-1 \times \frac{1}{2}\right)^2 = \frac{1}{4}$

(C) $\left(\frac{x}{y}\right)^2 = \left(\frac{-1}{\frac{1}{2}}\right)^2 = 4$: Maximum

(D) $\frac{x^2}{y} = \frac{(-1)^2}{\frac{1}{2}} = 2$

(E) $x^2y = (-1)^2 \times \frac{1}{2} = \frac{1}{2}$

205. Let the distance between the cities A and B be d miles.

Range of speeds of David = 20 miles per hour to 30 miles per hour.

Time taken by David to cover the distance = 5 hours.

So, the range of distance between cities A and B
= [20 × 5 miles] to [30 × 5 miles]

= 100 miles to 150 miles

=> $100 \le d \le 150 \ldots$ (i)

Range of speeds of Mark = 40 miles per hour to 60 miles per hour.

Time taken by Mark to cover the distance = 3 hours.

So, the range of distance between cities A and B
= [40 × 3 miles] to [60× 3 miles]

= 120 miles to 180 miles

=> $120 \le d \le 180 \ldots$ (ii)

Thus, from (i) and (ii), we have

Range of distance between cities A and B

= (Higher of the two minimum values) to (Lower of the two maximum values)

= (Higher of 100 and 120) to (Lower of 150 and 180)

= 120 miles to 150 miles

=> $120 \le d \le 150$

The only option that satisfies above is 135 miles.

The correct answer is Option B.

4.17 Co-ordinate geometry

206. The length of the line segment between 2 points $(x_1, y_1) = (-3, -6)$ and $(x_2, y_2) = (5, 0)$ is given by:

$$L = \sqrt{(x_2 - x_1)^2 + (y_2 - y_1)^2}$$

$$= \sqrt{(5 - (-3))^2 + (0 - (-6))^2}$$

$$= \sqrt{8^2 + 6^2}$$

$$= 10$$

Thus, the length of the diameter of the circle = 10.

Thus, the length of the radius of the circle = $\frac{10}{2} = 5$.

Thus, the area of the circle = $\pi \times 5^2 = 25\pi$.

The correct answer is Option C.

207. Equation of a line is: $y = mx + c$, where m is the slope and c is the Y-intercept.

Thus, we have the equation of the given line as:

$$y = 3x + 4$$

Let the required point on the above line be: $(a, 10)$.

Thus, we have

$$10 = 3a + 4$$

$$=> a = 2$$

Thus, the required X-coordinate of the point is 2.

The correct answer is Option A.

208. Equation of a line passing through a point (p, q) and slope m is given as:

$$y - q = m(x - p)$$

Thus, the equation of the line l that passes through the origin $(0, 0)$ and has slope 3 is:

$y - 0 = 3(x - 0)$

$=> y = 3x$

Since $(1, a)$ is a point on the line, we have

$a = 3 \times 1$

$=> a = 3 \ldots$ (i)

Since $(b, 2)$ is a point on the line, we have

$2 = 3 \times b$

$=> b = \frac{2}{3} \ldots$ (ii)

Thus, from (i) and (ii), we have

$\frac{a}{b} = \frac{3}{\left(\frac{2}{3}\right)} = \frac{9}{2}$

The correct answer is Option E.

209. The equation of a circle having center at (p, q) and radius r is:
$(x - p)^2 + (y - q)^2 = r^2$

Since the center of the circle is at $(3, 2)$, the equation of the circle is:
$(x - 3)^2 + (y - 2)^2 = r^2$

If a point (m, n) lies inside the circle $(x - p)^2 + (y - q)^2 = r^2$,
it must satisfy: $(m - p)^2 + (n - q)^2 < r^2$

Since $(-1,\ 2)$ lies inside the circle, it must satisfy:
$(-1 - 3)^2 + (2 - 2)^2 < r^2$

$=> r^2 > 16$

$=> r > 4 \ldots$ (i)

If a point (m, n) lies outside the circle, it must satisfy:
$(m - p)^2 + (n - q)^2 > r^2$

Since $(3, -4)$ lies outside the circle, it must satisfy:
$(3 - 3)^2 + (-4 - 2)^2 > r^2$

$=> r^2 < 36$

$=> -6 < r < 6$

Since r must be positive, we have

$0 < r < 6 \ldots$ (ii)

Thus, from (i) and (ii), we have

$4 < r < 6$

From the options, only $r = 5$ satisfies the above inequality.

The correct answer is Option A.

Alternate approach:

Since $(-1, 2)$ lies inside, the distance between $(-1, 2)$ and the center $(3, 2)$ is less than the radius.

Similarly, since $(3, -4)$ lies outside, the distance between $(3, -4)$ and the center $(3, 2)$ is greater than the radius. Thus, we have

$\sqrt{[(-1-3)^2 + (2-2)^2]} < r < \sqrt{[(3-3)^2 + (-4-2)^2]}$

$=> \sqrt{16} < r < \sqrt{36}$

$=> 4 < r < 6 => r = 5$ (only value among the options)

210. The three vertices are: (a, b), $(a, -b)$, and $(-a, -b)$

We know that $a < 0$ and $b > 0$

$=> -a > 0$ and $-b < 0$

We can see that the distance of each of the vertices from the origin $(0, 0) = \sqrt{a^2 + b^2}$

Thus, the three vertices are equidistant from the origin.

Alternately, we can see that the midpoint of the diagonal joining (a, b) and $(-a, -b)$ is $(0, 0)$, i.e. the origin.

Thus, the centre of the square is at the origin.

Thus, the square would be positioned as shown in the diagram below:

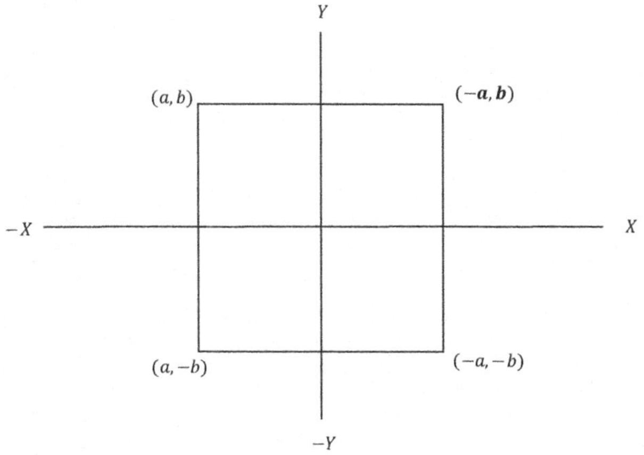

Thus, the fourth vertex would be $(-a, b)$, which lies in the first quadrant.

Thus, the X-coordinate of the fourth vertex is positive and the Y-coordinate of the fourth vertex is also positive.

Thus, the only point which also lies in the same quadrant is $(6, 2)$.

The correct answer is Option E.

211. Distance between two points (x_1, y_1) and (x_2, y_2) is given by $\sqrt{(x_2 - x_1)^2 + (y_2 - y_1)^2}$

Distance between the points $(0, 0)$ and $(5, 5)$

$= \sqrt{(5 - 0)^2 + (5 - 0)^2}$

$= \sqrt{50}$

$= 5\sqrt{2}$

Distance between the points $(0, 0)$ and $(10, 0)$

$= \sqrt{(10 - 0)^2 + (0 - 0)^2}$

$= \sqrt{100}$

$= 10$

Note: The distance is obviously 10 since (10, 0) is a point on the X-axis at a distance of 10 to the right of the origin.

Distance between the points $(10, 0)$ and $(5, 5)$

$= \sqrt{(10 - 5)^2 + (0 - 5)^2}$

$= \sqrt{50}$

$= 5\sqrt{2}$

Thus, the perimeter

$= 5\sqrt{2} + 10 + 5\sqrt{2}$

$= 10 + 10\sqrt{2}$

The correct answer is Option E.

212. Since the points are collinear, their slopes must be equal.

Slope of the line joining two points (x_1, y_1) and (x_2, y_2) is given by: $\left(\frac{y_2 - y_1}{x_2 - x_1}\right)$.

Slope of the line joining the points $(a, 0)$ and $(0, b)$

$= \frac{b - 0}{0 - a}$

$= -\frac{b}{a}$

Slope of the line joining the points $(a, 0)$ and $(1,\ 1)$

$= \frac{1 - 0}{1 - a}$

$= \frac{1}{1 - a}$

Thus, we have

$-\frac{b}{a} = \frac{1}{1 - a}$

$=> -b \times (1 - a) = a$

$=> -b + ab = a$

$=> b + a = ab$

$=> b = ab - a$

$=> b = a(b - 1)$

$=> a = \frac{b}{b - 1}$

The correct answer is Option C.

213. The area of the triangle formed by the line with the two axes
$= \frac{1}{2} \times$ (Length of X intercept) $\times$ (Length of Y intercept)

We have

$3y - 4x = 24 \ldots$ (i)

The graph of the line and the area of the triangle formed by the line and the axes is shown in the diagram below:

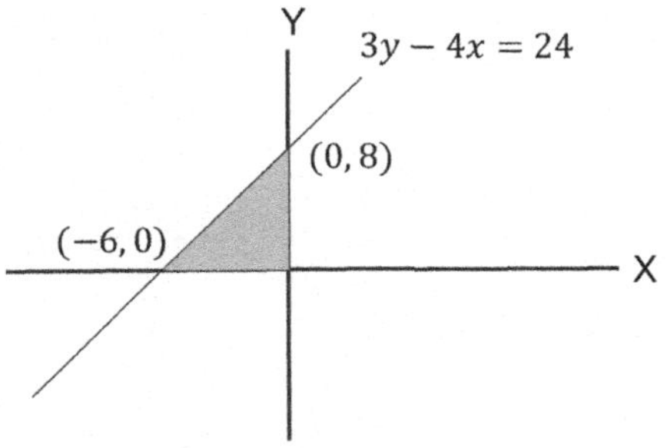

To calculate the X-intercept:

Substituting $y = 0$ in (i):

$x = -\frac{24}{4} = -6$

Thus, the length of the X-intercept = 6.

To calculate the Y-intercept:

Substituting $x = 0$ in (i):

$y = \frac{24}{3} = 8$

Thus, the length of the Y-intercept = 8.

Thus, required area

$= \frac{1}{2} \times 6 \times 8 = 24$

The correct answer is Option C.

Chapter 5

Talk to Us

Have a Question?

Email your questions to *info@manhattanreview.com*. We will be happy to answer you. Your questions can be related to a concept, an application of a concept, an explanation of a question, a suggestion for an alternate approach, or anything else you wish to ask regarding the GMAT Focus.

Please mention the page number when quoting from the book.

Best of luck!

Professor Dr. Joern Meissner
& The Manhattan Review Team